SPACE SCIENCE, EXPLORATION AND POLICIES

SPACE EXPLORATION

DEVELOPMENT, PLANNING AND CHALLENGES

SPACE SCIENCE, EXPLORATION AND POLICIES

Additional books and e-books in this series can be found on Nova's website under the Series tab.

SPACE SCIENCE, EXPLORATION AND POLICIES

SPACE EXPLORATION

DEVELOPMENT, PLANNING AND CHALLENGES

THOMAS R. REED
EDITOR

snova
New York

NOTICE TO THE READER

Library of Congress Cataloging-in-Publication Data

ISBN: 978-1-53615-032-2

Published by Nova Science Publishers, Inc. † New York

CONTENTS

Contents

PREFACE

Human spaceflight at NASA began in the 1960s with the Mercury and Gemini programs leading up to the Apollo moon landings. After the last lunar landing, Apollo 17, in 1972, NASA shifted its attention to low earth orbit operations with human spaceflight efforts that included the Space Shuttle and International Space Station programs through the remainder of the 20th century. Exploration of our solar system has brought great knowledge to our Nation's scientific and engineering community over the past several decades. As we expand our visions to explore new, more challenging destinations, we must also expand our technology base to support these new missions. NASA's Space Technology Mission Directorate is tasked with developing these technologies for future mission infusion and continues to seek answers to many existing technology gaps.

Chapter 1 - NASA is undertaking a trio of closely related programs to continue human space exploration beyond low-Earth orbit. All three programs (SLS, Orion, and EGS) are working toward a launch readiness date of no earlier than October 2019 for the first test flight. Each program is a complex technical and programmatic endeavor. Because all three programs must work together for launch, NASA must integrate the hardware and software from the separate programs into a working system capable of meeting its goals for deep space exploration. The House Committee on Appropriations report accompanying H.R. 2578 included a provision for

GAO to assess the progress of NASA's human space exploration programs. This chapter assesses (1) the benefits and challenges of NASA's approach for integrating these three programs and (2) the extent to which cross-program risks could affect launch readiness. GAO examined NASA policies, the results of design reviews, risk data, and other program documentation and interviewed NASA and other officials.

Chapter 2 - Exploration of our solar system has brought great knowledge to our Nation's scientific and engineering community over the past several decades. As we expand our visions to explore new, more challenging destinations, we must also expand our technology base to support these new missions. NASA's Space Technology Mission Directorate is tasked with developing these technologies for future mission infusion and continues to seek answers to many existing technology gaps. One such technology gap is related to compact power systems (>1 kWe) that provide abundant power for several years where solar energy is unavailable or inadequate. Below 1 kWe, Radioisotope Power Systems have been the workhorse for NASA and will continue, assuming its availability, to be used for lower power applications similar to the successful missions of Voyager, Ulysses, New Horizons, Cassini, and Curiosity. Above 1 kWe, fission power systems (FPSs) become an attractive technology offering a scalable modular design of the reactor, shield, power conversion, and heat transport subsystems. Near-term emphasis has been placed in the 1 to 10 kWe range that lies outside realistic radioisotope power levels and fills a promising technology gap capable of enabling both science and human exploration missions. History has shown that development of space reactors is technically, politically, and financially challenging and requires a new approach to their design and development. A small team of NASA and Department of Energy experts are providing a solution to these enabling FPS technologies starting with the lowest power and most cost-effective reactor series named "Kilopower" that is scalable from approximately 1 to 10 kWe.

Chapter 3 - NASA uses RPS to generate electrical power in missions in which solar panels or batteries would be ineffective. RPS convert heat generated by the radioactive decay of Pu-238 into electricity. DOE maintains a capability to produce RPS for NASA missions, as well as a limited and aging supply of Pu-238 that will be depleted in the 2020s, according to NASA and DOE officials and documentation. With NASA funding, DOE initiated the Pu-238 Supply Project in 2011, with a goal of producing 1.5 kg of new Pu-238 per year by 2026. Without new Pu-238, future NASA missions requiring RPS are at risk. GAO was asked to review planned RPS and Pu-238 production to support future NASA missions. This chapter (1) describes how NASA selects RPS for missions and what factors affect RPS and Pu-238 demand; and (2) evaluates DOE's progress and challenges in meeting NASA's RPS and Pu-238 demand. GAO reviewed NASA mission planning and DOE program documents, visited two DOE national laboratories involved in making new Pu-238 or RPS work, and interviewed agency officials.

Chapter 4 - This is an edited, reformatted and augmented accessible version of the United States Government Accountability Office Testimony Before the Subcommittee on Space, Committee on Science, Space, and Technology, House of Representatives, Publication No. GAO-18-161T, dated October 4, 2017.

In: Space Exploration
Editor: Thomas R. Reed

ISBN: 978-1-53615-032-2
© 2019 Nova Science Publishers, Inc.

Chapter 1

NASA HUMAN SPACE EXPLORATION: INTEGRATION APPROACH PRESENTS CHALLENGES TO OVERSIGHT AND INDEPENDENCE*

United States Government Accountability Office

ABBREVIATIONS

ASAP	Aerospace Safety Advisory Panel
CAIB	Columbia Accident Investigation Board
CDR	Critical Design Review
EGS	Exploration Ground Systems
EM-1	Exploration Mission 1
EM-2	Exploration Mission 2
EM-3	Exploration Mission 3

---

* This is an edited, reformatted and augmented version of United States Government Accountability Office; Report to Congressional Committees, Accessible Version, Publication No. GAO-18-28, dated October 2017.

EM-4	Exploration Mission 4
ESD	Exploration Systems Development
ESI	Exploration Systems Integration
IG	Inspector General
ITL	Integrated Test Laboratory
KDP	Key Decision Point
MDA	Missile Defense Agency
MDR	Mission Definition Review
NASA	National Aeronautics and Space Administration
Orion	Orion Multi-Purpose Crew Vehicle
PDR	Preliminary Design Review
S&MA	Safety and Mission Assurance
SDR	System Definition Review
SIR	System Integration Review
SLS	Space Launch System
SRR	System Requirements Review
V&V	Verification and Validation

WHY GAO DID THIS STUDY

NASA is undertaking a trio of closely related programs to continue human space exploration beyond low-Earth orbit. All three programs (SLS, Orion, and EGS) are working toward a launch readiness date of no earlier than October 2019 for the first test flight. Each program is a complex technical and programmatic endeavor. Because all three programs must work together for launch, NASA must integrate the hardware and software from the separate programs into a working system capable of meeting its goals for deep space exploration.

The House Committee on Appropriations report accompanying H.R. 2578 included a provision for GAO to assess the progress of NASA's human space exploration programs. This chapter assesses (1) the benefits and challenges of NASA's approach for integrating these three programs and (2) the extent to which cross-program risks could affect launch readiness. GAO examined NASA policies, the results of design reviews, risk data, and other program documentation and interviewed NASA and other officials.

WHAT GAO RECOMMENDS

Congress should consider directing NASA to establish baselines for SLS and EGS's missions beyond the first test flight. NASA's ESD organization should no longer dual-hat officials with programmatic and technical authority responsibilities. NASA partially concurred with our recommendation and plans to address it in the next year. But NASA did not address the need for the technical authority to be independent from programmatic responsibilities for cost and schedule. GAO continues to believe that this component of the recommendation is critical.

WHAT GAO FOUND

The approach that the National Aeronautics and Space Administration (NASA) is using to integrate its three human spaceflight programs into one system ready for launch offers some benefits, but it also introduces oversight challenges. To manage and integrate the three programs—the Space Launch System (SLS) vehicle; the Orion crew capsule; and supporting ground systems (EGS)— NASA's Exploration Systems Development (ESD) organization is using a more streamlined approach than has been used with other programs, and officials GAO spoke with believe that this approach provides cost savings and greater efficiency. However, GAO found two key challenges to the approach:

- The approach makes it difficult to assess progress against cost and schedule baselines. SLS and EGS are baselined only to the first test flight. In May 2014, GAO recommended that NASA baseline the programs' cost and schedule beyond the first test flight. NASA has not implemented these recommendations nor does it plan to; hence, it is contractually obligating billions of dollars for capabilities for the second flight and beyond without establishing baselines necessary to measure program performance.

- The approach has dual-hatted positions, with individuals in two programmatic engineering and safety roles also performing oversight of those areas. As the image below shows, this presents an environment of competing interests.

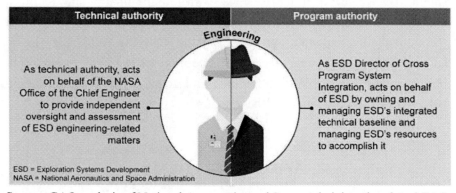
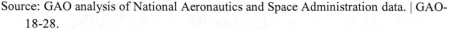

Source: GAO analysis of National Aeronautics and Space Administration data. | GAO-18-28.

Competing Interests between Engineering Technical Authority Role and Program Role.

These dual roles subject the technical authorities to cost and schedule pressures that potentially impair their independence. The Columbia Accident Investigation Board found in 2003 that this type of tenuous balance between programmatic and technical pressures was a contributing factor to that Space Shuttle accident.

NASA has lowered its overall cross-program risk posture over the past 2 years, but risk areas—related to software development and verification and validation, which are critical to ensuring the integrated body works as expected—remain. For example, delays and content deferral in Orion and SLS software development continue to affect ground systems software development and could delay launch readiness. GAO will continue to monitor these risks.

October 19, 2017

The Honorable Richard Shelby
Chairman

The Honorable Jeanne Shaheen
Ranking Member
Subcommittee on Commerce, Justice, Science, and Related Agencies
Committee on Appropriations
United States Senate

The Honorable John Culberson
Chairman

The Honorable José Serrano
Ranking Member
Subcommittee on Commerce, Justice, Science, and Related Agencies
Committee on Appropriations
House of Representatives

The National Aeronautics and Space Administration (NASA) is nearing the point when billions of dollars invested should begin to pay off with the first launch of systems needed to support deep space exploration by humans. This deep space exploration requires the capability to transport crew and large masses of cargo beyond low Earth orbit to distant destinations including the moon and eventually Mars. The Exploration Systems Development (ESD) organization within NASA's Human Exploration and Operations Mission Directorate is responsible for managing and integrating the three programs developing the specific capabilities needed.

- The Space Launch System (SLS) program is developing a vehicle to launch a crew capsule and cargo beyond low-Earth orbit.
- The Orion Multi-Purpose Crew Vehicle (Orion) program is developing a crew capsule to transport humans beyond low-Earth orbit.
- The Exploration Ground Systems (EGS) program is developing systems and infrastructure to support assembly, test, and launch of the SLS and Orion crew capsule, and recovery of the Orion crew capsule.

This portfolio of three programs is estimated to cost almost $24 billion—to include two Orion flights and one each for SLS and EGS—and constitute more than half of NASA's planned development budget. All three programs are necessary for the first integrated test flight, Exploration Mission 1 (EM-1), and are working to a launch readiness date of no earlier than October 2019.

NASA intends for ESD's portfolio of programs—SLS, Orion, and EGS—to provide an important capability for human exploration missions. Each of these programs represents a large, complex technical and programmatic endeavor. In addition, since all three programs must work together for launch, NASA faces the additional challenge of integrating the hardware and software from the separate programs into a working system capable of effectively meeting its goals for deep space exploration. Our prior work has shown that the integration and test phase often reveals unforeseen challenges leading to cost growth and schedule delays.[1]

GAO has designated NASA's management of acquisitions as a high-risk area for more than two decades. In February 2017, we found that the agency has continued to make progress in reducing risk on major projects after

[1] GAO, *Space Launch System: Resources Need to be Matched to Requirements to Decrease Risk and Support Long Term Affordability*, GAO-14-631 (Washington, D.C.: July 23, 2014); *Space Launch System: Management Tools Should Better Track to Cost and Schedule Commitments to Adequately Monitor Increasing Risk*, GAO-15-596 (Washington, D.C.: July 16, 2015); and *James Webb Space Telescope: Project on Track but May Benefit from Improved Contractor Data to Better Understand Costs*, GAO-16-112 (Washington, D.C.: Dec. 17, 2015).

previously struggling with poor cost estimation, weak oversight, and risk underestimation. We also found that the Orion, SLS, and EGS programs are generally better positioned for success than past crewed vehicle efforts that were canceled after facing acquisitions problems and funding-related issues. Nevertheless, as we have reported, management weaknesses—including overly ambitious schedules, unreliable cost estimating, limited reserves, and operating for extended periods of time without definitized contracts—have increased the likelihood that the programs will incur schedule delays and cost overruns, particularly when coupled with the technical risks that are inherent in any human spaceflight development.[2] In April 2017, we found that it was unlikely that the ESD programs would achieve the planned November 2018 launch readiness date and recommended that NASA reassess the date. NASA agreed with this recommendation and stated that it would establish a new launch readiness date in fall 2017.[3] Subsequently, in June 2017, NASA sent notification to Congress that EM-1's recommended launch date would be no earlier than October 2019.

The House Committee on Appropriations included a provision in its 2015 report for GAO to review the acquisition progress of NASA's human exploration programs, including Orion, SLS, and EGS.[4] This chapter is the latest in a series of reports addressing the mandate. This chapter assesses (1) the benefits and challenges of NASA's approach for integrating and assessing the programmatic and technical readiness of Orion, SLS, and EGS; and (2) the extent to which ESD is managing cross-program risks that could affect launch readiness.

To assess the benefits and challenges of NASA's approach for integration, we obtained and analyzed NASA program policies governing program and technical integration, including cost, schedule, and risk. We obtained and analyzed ESD implementation plans to assess the role of ESD in cross program integration of SLS, Orion, and EGS and reviewed briefings explaining ESD's approach to programmatic and technical integration,

---

[2] GAO, *High Risk Series: Progress on Many High-Risk Areas, While Substantial Efforts Needed on Others*, GAO-17-317 (Washington, D.C.: Feb. 15, 2017).

[3] GAO, *NASA Human Space Exploration: Delay Likely for First Exploration Mission*, GAO-17-414 (Washington, D.C.: Apr. 27, 2017).

[4] H.R. Rep. No. 114-130, at 60-61 (2015), accompanying H.R. 2578.

including implementation of systems engineering and integration. In addition, we assessed the scope of NASA's funding estimates for the second exploration mission and beyond against best practices criteria outlined in GAO's cost estimating guidebook.[5] We reviewed the 2003 Columbia Accident Investigation Board Report's findings and recommendations related to culture and organizational management of human spaceflight programs as well as the Constellation program's lessons learned report. We met with the technical authorities and other representatives from the NASA Office of the Chief Engineer, Office of Safety and Mission Assurance, Crew Health and Safety, and addressed cost and budgeting issues with the Chief Financial Officer, and discussed and documented their roles in executing and overseeing the ESD programs. We also interviewed outside subject matter experts to gain their insight of ESD's implementation of NASA's program management policies on the independent technical authority structure.

To assess the extent to which ESD is managing cross-program risks that could affect launch readiness, we obtained and reviewed NASA and ESD risk management policies, detailed monthly and quarterly briefings and documentation from Cross-Program Systems Integration and Programmatic and Strategic Integration teams explaining ESD's approach to identifying, tracking, and mitigating cross-program risks. We conducted an analysis of ESD's risk dataset and the programs' detailed risk reports which list program risks and their potential schedule impacts, including mitigation efforts to date. We examined risk report data from Design to Synchronization (Design to Sync) to Build to Synchronization (Build to Sync) and focused our analyses to identify risks with current mitigation plans to determine if risk mitigation plans are proceeding on schedule. We supplemented this analysis with interviews of responsible ESD officials. For more information on our scope and methodology, see appendix I.

We conducted this performance audit from August 2016 to October 2017 in accordance with generally accepted government auditing standards. Those standards require that we plan and perform the audit to obtain

---

[5] GAO, *GAO Cost Estimating and Assessment Guide: Best Practices for Developing and Managing Capital Program Costs,* GAO-09-3SP (Washington, D.C.: March 2009).

sufficient, appropriate evidence to provide a reasonable basis for our findings and conclusions based on our audit objectives. We believe that the evidence obtained provides a reasonable basis for our findings and conclusions based on our audit objectives.

BACKGROUND

Human spaceflight at NASA began in the 1960s with the Mercury and Gemini programs leading up to the Apollo moon landings. After the last lunar landing, Apollo 17, in 1972, NASA shifted its attention to low earth orbit operations with human spaceflight efforts that included the Space Shuttle and International Space Station programs through the remainder of the 20th century. In the early 2000s, NASA once again turned its attention to cislunar and deep space destinations, and in 2005 initiated the Constellation program, a human exploration program that was intended to be the successor to the Space Shuttle.[6] The Constellation program was canceled, however, in 2010 due to factors that included cost and schedule growth and funding gaps.

Following Constellation, the National Aeronautics and Space Administration Authorization Act of 2010 directed NASA to develop a Space Launch System, to continue development of a crew vehicle, and prepare infrastructure at Kennedy Space Center to enable processing and launch of the launch system.[7] To fulfill this direction, NASA formally established the SLS program in 2011. Then, in 2012, the Orion project transitioned from its development under the Constellation program to a new development program aligned with SLS. To transition Orion from Constellation, NASA adapted the requirements from the former Orion plan with those of the newly created SLS and the associated ground systems programs. In addition, NASA and the European Space Agency agreed that it would provide a portion of the service module for Orion. Figure 1 provides

---

[6] Cislunar is the area between earth and the moon. Deep space encompasses the rest of the solar system.
[7] Pub. L. No. 111-267, § 302, 303, and 305.

details about the heritage of each SLS hardware element and its source as well as identifies the major portions of the Orion crew vehicle.

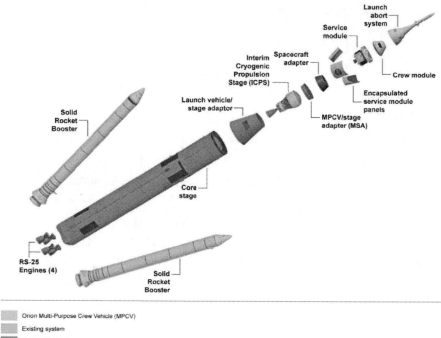

Source: GAO analysis of National Aeronautics and Space Administration data (data and images). | GAO-18-28.

Figure 1. Space Launch System and Orion Multi-Purpose Crew Vehicle Hardware.

The EGS program was established to modernize the Kennedy Space Center to prepare for integrating hardware from the three programs as well as processing and launching SLS and Orion and recovery of the Orion crew capsule. EGS is made up of nine major components, including: the Vehicle Assembly Building, Mobile Launcher, Launch Control Center and software, Launch Pad 39B, Crawler-Transporter, Launch Equipment Test Facility, Spacecraft Offline Processing, Launch Vehicle Offline Processing, and Landing and Recovery. See Figure 2 for pictures of the Mobile Launcher, Vehicle Assembly Building, Launch Pad 39B, and Crawler-Transporter.

Source: National Aeronautics and Space Administration. | GAO-18-28.

Figure 2. Select Components of Exploration Ground Systems Program.

NASA's Exploration Systems Development (ESD) organization is responsible for directing development of the three individual human spaceflight programs—SLS, Orion, and EGS—into a human space exploration system. The integration of these programs is key because all three systems must work together for a successful launch. The integration activities for ESD's portfolio occur at two levels in parallel throughout the life of the programs: as individual efforts to integrate the various elements managed within the separate programs and as a joint effort to integrate the three programs into an exploration system.

The three ESD programs support NASA's long term goal of sending humans to distant destinations, including Mars. NASA's approach to developing and demonstrating the technologies and capabilities to support their long term plans for a crewed mission to Mars includes three general stages of activities—*Earth Reliant, Proving Ground, and Earth Independent.*

- Earth Reliant: From 2016 to 2024, NASA's planned exploration is focused on research aboard the International Space Station. On the International Space Station, NASA is testing technologies and advancing human health and performance research that will enable deep space, long duration missions.
- Proving Ground: From the mid-2020s to early-2030s, NASA plans to learn to conduct complex operations in a deep space environment that allows crews to return to Earth in a matter of days. Primarily operating in cislunar space—the volume of space around the moon featuring multiple possible stable staging orbits for future deep space missions—NASA will advance and validate capabilities required for humans to live and work at distances much farther away from our home planet, such as on Mars.
- Earth Independent: From the early-2030s to the mid-2040s, planned activities will build on what NASA learns on the space station and in deep space to enable human missions to the vicinity of Mars, possibly to low-Mars orbit or one of the Martian moons, and eventually the Martian surface.

The first launch of the integrated ESD systems, EM-1, is a Proving Ground mission. EM-1 is planned as an uncrewed test flight currently planned for no earlier than October 2019 that will fly about 70,000 kilometers beyond the moon. The second launch, Exploration Mission 2 (EM-2), which will utilize an evolved SLS variant with a more capable upper stage, is also a Proving Ground mission planned for no later than April 2023. EM-2 is expected to be a 10- to 14-day crewed flight with up to four astronauts that will orbit the moon and return to Earth to demonstrate the baseline Orion vehicle capability. NASA eventually plans to develop larger and more capable versions of the SLS to support Proving Ground and Earth Independent missions after EM-2.[8]

As noted above, in April 2017 we found that given the combined effects of ongoing technical challenges in conjunction with limited cost and schedule reserves, it was unlikely that the ESD programs would achieve the November 2018 launch readiness date. We recommended that NASA confirm whether the EM-1 launch readiness date of November 2018 was achievable, as soon as practicable but no later than as part of its fiscal year 2018 budget submission process. We also recommended that NASA propose a new, more realistic EM-1 date if warranted. NASA agreed with both recommendations and stated that it was no longer in its best interest to pursue the November 2018 launch readiness date. Further, NASA stated that, in fall 2017, it planned to establish a new launch readiness date.[9] Subsequently, in June 2017, NASA sent notification to Congress that EM-1's recommended launch date would be no earlier than October 2019.

The life cycle for NASA space flight projects consists of two phases— formulation, which takes a project from concept to preliminary design, and implementation, which includes building, launching, and operating the system, among other activities. NASA further divides formulation and implementation into pre-phase A through phase F. Major projects must get approval from senior NASA officials at key decision points before they can enter each new phase. The three ESD programs are completing design and fabrication efforts prior to beginning Phase D system assembly, integration and test, launch and checkout. Figure 3 depicts NASA's life cycle for space flight projects.

---

[8] ESD officials indicated that moving forward NASA intends to replace the Earth Reliant, Proving Ground, Earth Independent planning framework with a new planning framework called Deep Space Gateway. Under this new framework, NASA anticipates a first phase of exploration near the moon using current technologies that will allow NASA to gain experience with extended operations farther from Earth than previously completed. According to NASA, these missions will enable it to develop new techniques and apply innovative approaches to solving problems in preparation for longer-duration missions far from Earth.

[9] GAO-17-414.

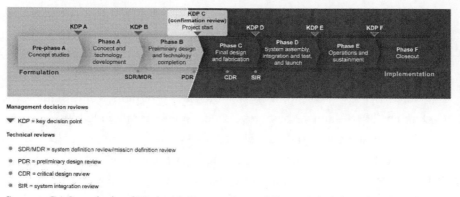

Management decision reviews

▼ KDP = key decision point

Technical reviews

✳ SDR/MDR = system definition review/mission definition review

✳ PDR = preliminary design review

✳ CDR = critical design review

✳ SIR = system integration review

Source: GAO analysis of National Aeronautics and Space Administration data. | GAO-
    18-28.

Figure 3. NASA's Life Cycle for Space Flight Projects.

NASA'S INTEGRATION APPROACH OFFERS SOME BENEFITS BUT COMPLICATES OVERSIGHT AND IMPAIRS INDEPENDENCE

NASA's approach for integrating and assessing programmatic and technical readiness, executed by ESD, differs from prior NASA human spaceflight programs. This new approach offers some cost and potential efficiency benefits. However, it also brings challenges specific to its structure. In particular, there are oversight challenges because only one of the three programs, Orion, has a cost and schedule estimate for EM-2. NASA is already contractually obligating money on SLS and EGS for EM-2, but the lack of cost and schedule baselines for these programs will make it difficult to assess progress over time. Additionally, the approach creates an environment of competing interests because it relies on dual-hatted staff to manage technical and safety aspects on behalf of ESD while also serving as independent oversight of those same areas.

Integration Approach Differs from Past Human Spaceflight Programs

NASA is managing the human spaceflight effort differently than it has in the past. Historically, NASA used a central management structure to manage human spaceflight efforts for the Space Shuttle and the Constellation programs. For example, both the Shuttle and Constellation programs were organized under a single program manager and used a contractor to support integration efforts. Additionally, the Constellation program was part of a three-level organization—the Exploration Systems Mission Directorate within NASA headquarters, the Constellation program, and then projects, including the launch vehicle, crew capsule, ground systems, and other lunar-focused projects, managed under the umbrella of Constellation. Figure 4 illustrates the three-level structure used in the Constellation program.

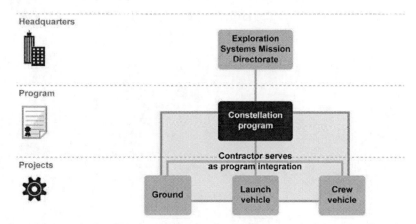

Source: GAO analysis of National Aeronautics and Space Administration data. | GAO-18-28.

Figure 4. Constellation Used Three-Level Organizational Structure.

In the Constellation program, the programmatic workforce was distributed within the program and projects. For example, systems engineering and integration organizations—those offices responsible for making separate technical designs, analyses, organizations and hardware come together to deliver a complete functioning system—were embedded within both the Constellation program and within each of the projects.

NASA's current approach is organized with ESD, rather than a contractor, as the overarching integrator for the three separate human spaceflight programs—SLS, Orion, and EGS. ESD manages both the programmatic and technical cross-program integration, and primarily relies on personnel within each program to implement its integration efforts. Exploration Systems Integration, an office within ESD, leads the integration effort from NASA headquarters. ESD officials stated that this approach is similar to that used by the Apollo program, wherein the program was also managed out of NASA headquarters.[10] Within Exploration Systems Integration, the Cross-Program Systems Integration sub-office is responsible for technical integration and the Programmatic and Strategic Integration sub-office is responsible for integrating the financial, schedule, risk management, and other programmatic activities of the three programs. The three programs themselves perform the hardware and software integration activities. This organizational structure that consists of two levels is shown in Figure 5.

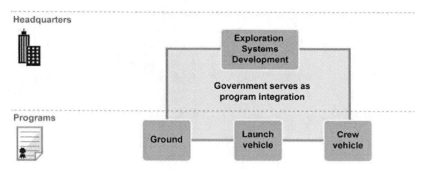

Source: GAO analysis of National Aeronautics and Space Administration data. | GAO-18-28.

Figure 5. Exploration Systems Development Organization's Approach Uses a Two-Level Organizational Structure.

---

[10] ESD officials indicated that the Space Shuttle program systems engineering and integration was also managed out of NASA headquarters for a short time after the Challenger accident in 1986.

ESD is executing a series of six unique integration-focused programmatic and technical reviews at key points within NASA's acquisition life cycle, as shown in figure 6, to assess whether NASA cost, schedule, and technical commitments are being met for the three-program enterprise.

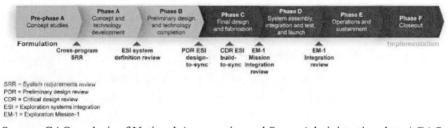

Source: GAO analysis of National Aeronautics and Space Administration data. | GAO-18-28.

Figure 6. Exploration Systems Development Organization's Integration Reviews.

These reviews cover the life cycle of the integrated programs to EM-1, from formulation to readiness to launch. Some of these reviews are unique to ESD's role as integration manager, For example, ESD established two checkpoints—Design to Sync in 2015 and Build to Sync in 2016. The purpose of Design to Sync was to assess the ability of the integrated preliminary design to meet system requirements, similar to a preliminary design review and the purpose of Build to Sync was to assess the maturity of the integrated design in readiness for assembly, integration, and test, similar to a critical design review (CDR).[11] At both events, NASA assessed the designs as ready to proceed. Key participants in these integration reviews include ESD program personnel and the Cross-Program Systems Integration and Programmatic and Strategic Integration staff that are responsible for producing and managing the integration activities.

---

[11] Within NASA, the preliminary design review demonstrates that the preliminary design meets all system requirements with acceptable risk and within the cost and schedule constraints and establishes the basis for proceeding with detailed design. The CDR demonstrates that the maturity of the design is appropriate to support proceeding with full-scale fabrication, assembly, integration, and test. CDR determines that the technical effort is on track to complete the system development, meeting performance requirements within the identified cost and schedule constraints.

ESD's Integration Approach Offers Some Cost Avoidance and Potential Efficiency Gains

ESD's integration approach offers some benefits in terms of cost avoidance relative to NASA's most recent human spaceflight effort, the Constellation program. NASA estimated it would need $190 million per year for the Constellation program integration budget. By comparison, between fiscal years 2012 and 2017, NASA requested an average of about $84 million per year for the combined integration budgets of the Orion, SLS, EGS, and ESD. This combined average of about $84 million per year represents a significant decrease from the expected integration budget of $190 million per year under the Constellation program. In addition, as figure 7 shows, NASA's initial estimates for ESD's required budget for integration are close to the actuals for fiscal years 2012-2017. NASA originally estimated that ESD's budget for integration would require approximately $30 million per year. ESD's integration budget was less than $30 million in fiscal years 2012 and 2013 and increased to about $40 million in fiscal year 2017—an average of about $30 million a year.

According to NASA officials, some of the cost avoidance can be attributed to the difference in workforce size. The Constellation program's systems engineering and integration workforce was about 800 people in 2009, the last full year of the program; whereas ESD's total systems engineering and integration workforce in 2017 was about 500 people, including staff resident in the individual programs.

ESD officials also stated that, in addition to cost avoidance, their approach provides greater efficiency. For example, ESD officials said that decision making is much more efficient in the two-level ESD organization than Constellation's three-level organization because the chain of command required to make decisions is shorter and more direct. ESD officials also indicated that the post-Constellation elimination of redundant systems engineering and integration staff at program and project levels contributed to efficiency. Additionally, they stated that program staff are invested in both their respective programs and the integrated system because they work on behalf of the programs and on integration issues for ESD. Finally, they said another contribution to increased efficiency was NASA's decision to establish SLS, Orion, and EGS as separate programs, which allowed each program to proceed at its own pace.

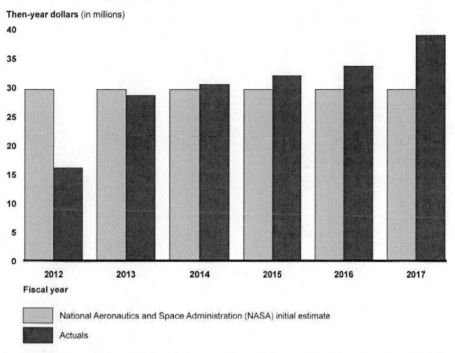

Then-year dollars (in millions)

Source: GAO analysis of National Aeronautics and Space Administration data. | GAO-18-28.

Figure 7. Exploration Systems Development Organization's Integration Budget Fiscal Years 2012-2017.

One caveat to this benefit, however, is that ESD's leaner organization is likely to face challenges to its efficiency in the integration and test phases of the SLS, Orion, and EGS programs. We analyzed the rate at which ESD has reviewed and approved the different types of launch operations and ground processing configuration management records for integrated SLS, Orion, and EGS operations, and found that the process is proceeding more slowly than ESD anticipated. For example, as figure 8 illustrates, ESD approved 403 fewer configuration management records than originally planned in the period from March 2016 through June 2017. According to an ESD official, the lower-than-planned approval rate resulted from the time necessary to establish and implement a new review process as well as final records being slower to arrive from the programs for review than ESD anticipated. Additionally, the official stated that the records required differing review timelines because they varied in size and scope.

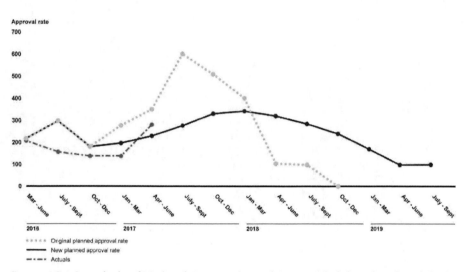

Source: GAO analysis of National Aeronautics and Space Administration data. | GAO-18-28.

Figure 8. Exploration Systems Development Organization's Configuration Management Records Approval Rate.

As figure 8 shows, ESD originally expected the number of items that needed review and approval to increase and create a "bow wave" during 2017 and 2018. In spring 2017, ESD re-planned its review and approval process and flattened the bow wave. The final date for review completion is now aligned with the new planned launch readiness date of no earlier than October 2019, which added an extra year to ESD's timeframe to complete the record reviews. While the bow wave is not as steep as it was under the original plan, ESD will continue to have a large number of records that require approval in order to support the launch readiness date. An ESD official stated that NASA had gained experience managing such a bow wave as it prepared for Orion's 2014 exploration flight test launch aboard a Delta

IV rocket and as part of the Constellation program's prototype Ares launch in 2009, but acknowledged that ESD will need to be cautious that its leaner staff is not overwhelmed with documentation, which could slow down the review process.

ESD's Approach Complicates Oversight Because There Is No Mechanism to Assess Affordability beyond First Mission

ESD is responsible for overall affordability for SLS, Orion, and EGS, while each of the programs develops and maintains an individual cost and schedule baseline. The baseline is created at the point when a program receives NASA management approval to proceed into final design and production. In their respective baselines, as shown in Table 1, SLS and EGS cost and schedule are baselined to EM-1, and Orion's are baselined to EM-2. NASA documentation indicates that Orion's baselines are tied to EM-2 because that is the first point at which it will fulfill its purpose of carrying crew. Should NASA determine it is likely to exceed its cost estimate baseline by 15 percent or miss a milestone by 6 months or more, NASA is required to report those increases and delays—along with their impacts—to Congress. In June 2017, NASA sent notification to Congress that the schedule for EM-1 has slipped beyond the allowed 6- month threshold, but stated that cost is expected to remain within the 15 percent threshold.[12]

NASA has not established EM-2 cost baselines or expected total life-cycle costs for SLS and EGS, including costs related to the larger and more capable versions of SLS needed to implement the agency's plans to send crewed missions to Mars. GAO's *Cost Estimating and Assessment Guide*, a guidebook of cost estimating best practices developed in concert with the public and private sectors, identifies baselines as a critical means for

---

[12] In 2005, Congress required NASA to report cost and schedule baselines—benchmarks against which changes can be measured—for all programs and projects with estimated life-cycle costs of at least $250 million that have been approved to proceed to implementation. Congress also required NASA to report to it when development cost growth or schedule delays exceeded certain thresholds. National Aeronautics and Space Administration Authorization Act of 2005, Pub. L. No. 109-155, § 103; 51 U.S.C. § 30104.

measuring program performance over time and addresses how a baseline backed by a realistic cost estimate increases the probability of a program's success.[13]

Table 1. Exploration Systems Development Organization-Managed Human Exploration Programs Are Baselined to Different Missions

Exploration Systems Development Human Exploration Programs	Cost baseline (Then-year dollars)	Baselined launch readiness date	Revised launch readiness date	Mission
Space Launch System	9.7 billion	November 2018	No earlier than October 2019	Exploration Mission 1
Exploration Ground Systems	2.8 billion	November 2018	No earlier than October 2019	Exploration Mission 1
Orion Multi-Purpose Crew Vehicle	11.3 billion	April 2023	not applicable	Exploration Mission 2

Source: GAO analysis of National Aeronautics and Space Administration data. | GAO-18-28.

In addition, prior GAO work offers insight into the benefits of how baselines enhance a program's transparency. For example, we found in 2009 that costs for the Missile Defense Agency's (MDA) ballistic missile defense system had grown by at least $1 billion, and that lack of baselines for each block of capability hampered efforts to measure progress and limited congressional oversight of MDA's work.[14] MDA responded to our recommendation to establish these baselines and, in 2011, we reported that MDA had a new process for setting detailed baselines, which had resulted in a progress report to Congress more comprehensive than the one it provided in 2009.[15]

To that end, we have made recommendations in the past on the need for NASA to baseline the programs' costs for capabilities beyond EM-1; however, a significant amount of time has passed without NASA taking

---

[13] GAO-09-3SP.

[14] GAO, *Defense Acquisitions: Production and Fielding of Missile Defense Components Continue with Less Testing and Validation Than Planned*, GAO-09-338 (Washington, D.C.: Mar. 13, 2009).

[15] GAO, *Missile Defense: Actions Needed to Improve Transparency and Accountability*, GAO-11-372 (Washington, D.C.: Mar. 24, 2011).

steps to fully implement these recommendations. Specifically, in May 2014, we recommended that, to provide Congress with the necessary insight into program affordability, ensure its ability to effectively monitor total program costs and execution, and to facilitate investment decisions, NASA's Administrator should direct the Human Exploration and Operations Mission Directorate to:

- Establish a separate cost and schedule baseline for work required to support the SLS for EM-2 and report this information to the Congress through NASA's annual budget submission. If NASA decides to fly the SLS configuration used in EM-2 beyond EM-2, establish separate life cycle cost and schedule baseline estimates for those efforts, to include funding for operations and sustainment, and report this information annually to Congress via the agency's budget submission; and

- Establish separate cost and schedule baselines for each additional capability that encompass all life cycle costs, to include operations and sustainment, because NASA intends to use the increased capabilities of the SLS, Orion, and ground support efforts well into the future and has chosen to estimate costs associated with achieving the capabilities.

As part of the latter recommendation, we stated that, when NASA could not fully specify costs due to lack of well-defined missions or flight manifests, the agency instead should forecast a cost estimate range—including life cycle costs—having minimum and maximum boundaries and report these baselines or ranges annually to Congress via the agency's budget submission.[16]

In its comments on our 2014 report, NASA partially concurred with these two recommendations, noting that much of what it had already done or expected to do would address them. For example, the agency stated that establishing the three programs as separate efforts with individual cost and

---

[16] GAO, *NASA: Actions Needed to Improve Transparency and Assess Long-Term Affordability of Human Exploration Programs*, GAO-14-385 (Washington, D.C.: May 8, 2014).

schedule commitments met GAO's intent as would its plans to track and report development, operations, and sustainment costs in its budget to Congress as the capabilities evolved. In our response, we stated that while NASA's prior establishment of three separate programs lends some insight into expected costs and schedule at the broader program level, it does not meet the intent of the two recommendations because cost and schedule identified at that level is unlikely to provide the detail necessary to monitor the progress of each block against a baseline. Further, reporting the costs via the budget process alone will not provide information about potential costs over the long term because budget requests neither offer all the same information as life-cycle cost estimates nor serve the same purpose. Life-cycle cost estimates establish a full accounting of all program costs for planning, procurement, operations and maintenance, and disposal and provide a long-term means to measure progress over a program's life span.

In 2016, NASA requested closure of these recommendations, citing, among other factors, changes to the programs' requirements, design, architecture, and concept of operations. However, NASA's request did not identify any steps taken to meet the intent of these two recommendations, such as establishing cost and schedule baselines for EM-2, baselines for each increment of SLS, Orion, or ground systems capability, or documentation of life cycle cost estimates with minimum and maximum boundaries. Further, a senior level ESD official told us that NASA does not intend to establish a baseline for EM-2 because it is not required to do so. The limited scope that NASA has chosen to use as the basis for formulating the programs' cost baselines does not provide the transparency necessary to assess long-term affordability. Plainly, progress cannot be assessed without a baseline that serves as a means to compare current costs against expected costs; consequently, it becomes difficult to assess program affordability and for Congress to make informed budgetary decisions.

NASA's lack of action in regards to our 2014 recommendations means that it is now contractually obligating NASA to spend billions of dollars in potential costs for EM-2 and beyond without a baseline against which to assess progress. For example:

- in fiscal year 2016, the SLS program awarded two contracts to Aerojet Rocketdyne: a $175 million contract for RL-10 engines to power the exploration upper stage during EM-2 and EM-3 and a $1.2 billion contract to restart the RS-25 production line required for engines for use beyond EM-4, and to produce at least 4 additional RS-25 engines;[17]
- in 2017, SLS modified the existing Boeing contract upwards by $962 million for work on the exploration upper stage that SLS will use during EM-2 and future flights; and
- on a smaller scale, in fiscal year 2016 the EGS program obligated $4.8 million to support the exploration upper stage and EM-2.

As illustrated by these contracting activities, the SLS program is obligating more funds for activities beyond EM-1 than Congress directed. Specifically, of approximately $2 billion appropriated for the SLS program, the Consolidated Appropriations Act, 2016 directed that NASA spend not less than $85 million for enhanced upper stage development for EM-2.[18] NASA has chosen to allocate about $360 million of its fiscal year 2016 SLS appropriations towards EM-2, including enhanced upper stage development, additional performance upgrades, and payload adapters, without a baseline to measure progress and ensure transparency. The NASA Inspector General (IG) also recently reported that NASA is spending funds on EM-2 efforts without a baseline in place and expressed concerns about the need for EM-2 cost estimates.[19] Because NASA has not implemented our recommendations, it may now be appropriate for Congress to take action to require EM-2 cost and schedule baselines for SLS and EGS, and separate cost and schedule baselines for additional capabilities developed for Orion, SLS, and EGS for missions beyond EM-2. These baselines would be important tools for Congress to make informed, long-term budgetary

---

[17] The RS-25 was the Space Shuttle's main engine. The SLS program is using a modified RS-25 to power the SLS core stage.

[18] Pub. L. No. 114-113 (2015), 129 Stat. 2316.

[19] NASA, Office of Inspector General, Office of Audits, *NASA's Plans for Human Exploration Beyond Low Earth Orbit*, (Washington, D.C.: April 13, 2017).

decisions with respect to NASA's future exploration missions, including Mars.

Organizational Structure Impairs Independence of Engineering and Safety Technical Oversight

NASA's governance model prescribes a management structure that employs checks and balances among key organizations to ensure that decisions have the benefit of different points of view and are not made in isolation. As part of this structure, NASA established the technical authority process as a system of checks and balances to provide independent oversight of programs and projects in support of safety and mission success through the selection of specific individuals with delegated levels of authority. The technical authority process has been used in other parts of the government for acquisitions, including the Department of Defense and Department of Homeland Security. ESD is organizationally connected to three technical authorities within NASA.

- The Office of the Chief Engineer technical authority is responsible for ensuring from an independent standpoint that the ESD engineering work meets NASA standards,
- The Office of Safety and Mission Assurance technical authority is responsible for ensuring from an independent standpoint that ESD products and processes satisfy NASA's safety, reliability, and mission assurance policies, and
- The Office of Chief Health and Medical technical authority is responsible for ensuring from an independent standpoint that ESD programs meet NASA's health and medical standards.

These NASA technical authorities have delegated responsibility for their respective technical authority functions directly to ESD staff.[20] According to NASA's project management requirements, the program or project manager is ultimately responsible for the safe conduct and successful outcome of the program or project in conformance with governing requirements and those responsibilities are not diminished by the implementation of technical authority.

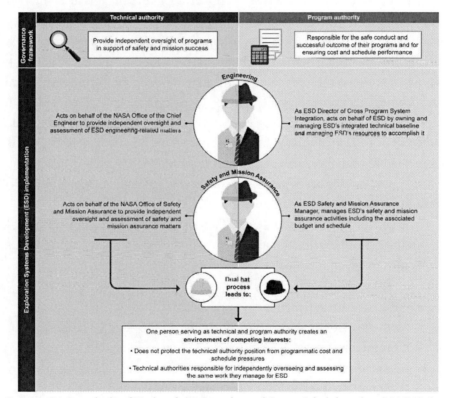

Source: GAO analysis of National Aeronautics and Space Administration (NASA) data. | GAO-18-28.

Figure 9. Conflicting Roles and Responsibilities of Exploration Systems Development Organization's Engineering and Safety and Mission Assurance Technical Authorities.

---

[20] NASA officials indicated that for most other NASA programs, technical authority is delegated to the program level through the Office of the Director for the NASA center where the program is executed.

ESD has established an organizational structure in which the technical authorities for engineering and safety and mission assurance (S&MA) are dual hatted to also serve simultaneously in programmatic positions. The chief engineer technical authority also serves as the Director of ESD's Cross Program System Integration Office and the S&MA technical authority also serves as the ESD Safety and Mission Assurance Manager. In their programmatic roles for ESD, the individuals manage resources, including budget and schedule, to address engineering and safety issues. In their technical authority roles, these same individuals are to provide independent oversight of programs and projects in support of safety and mission success. Having the same individual simultaneously fill both a technical authority role and a program role creates an environment of competing interests where the technical authority may be subject to impairments in their ability to impartially and objectively assess the programs while at the same time acting on behalf of ESD in programmatic capacities. This duality makes them more subject to program pressures of cost and schedule in their technical authority roles. Figure 9 describes some of the conflicting roles and responsibilities of these officials in their two different positions.

The concurrency of duties leaves the positions open to conflicting goals of safety, cost, and schedule and increases the potential for the technical authorities to become subject to cost and schedule pressures. For example:

- the dual-hatted engineering and S&MA technical authorities serve on decision-making boards both in technical authority and programmatic capacities, making them responsible for providing input on technical and safety decisions while also keeping an eye on the bottom line for ESD's cost and schedule; and
- the technical authorities are positioned such that they have been the reviewers of the ESD programmatic areas they manage—in essence, "grading their own homework." For example, at ESD's Build to Sync review in 2016, the engineering and S&MA technical authorities evaluated the areas that they manage in their respective capacities as ESD Director of Cross Program System Integration and ESD Safety and Mission Assurance Manager. This process

relied on their abilities as individuals to completely separate the two hats—using one hand to put on the ESD hat and manage technical and safety issues within programmatic cost and schedule constraints and using the other hand to take off that hat and assess the same issues with an independent eye.

NASA officials identified several reasons why the dual-hat structure works for their purposes. Agency officials stated that one critical factor to successful dual-hatting is having the "right" people in those dual-hat positions—that is, personnel with the appropriate technical knowledge to do the work and the ability to act both on behalf of ESD and independent of it. Officials also indicated that technical authorities retain independence because their technical authority reporting paths and performance reviews are all within their technical authority chain of command rather than under the purview of the ESD chain of command.

Additionally, agency officials said that dual-hat roles are a commonplace practice at NASA and cited other factors in support of the approach, including that:

- it would not be an efficient use of resources to have an independent technical authority with no program responsibilities because that person would be unlikely to have sufficient program knowledge to provide useful insight and could slow the program's progress;
- a technical authority that does not consider cost and schedule is not helpful to the program because it is unrealistic to disregard those aspects of program management;
- a strong dissenting opinion process is in place and allows for issues to be raised through various levels to the Administrator level within NASA; and
- ESD receives additional independent oversight through three NASA internal organizations—the independent review teams that provide independent assessments of a program's technical and programmatic status and health at key points in its life cycle; the NASA Engineering and Safety Center that conducts independent

safety and mission success-related testing, analysis, and
assessments of NASA's high-risk projects; and the Aerospace
Safety Advisory Panel (ASAP) that independently oversees
NASA's safety performance.

These factors that NASA officials cite in support of the dual-hat
approach minimize the importance of having independent oversight and
place ESD at risk of fostering an environment in which there is no longer a
balance between preserving safety with the demands of maintaining cost and
schedule. The Columbia Accident Investigation Board (CAIB) report—the
result of an in-depth assessment of the technical and organizational causes
of the Columbia accident—concluded that NASA's organization for the
Shuttle program combined, among other things, all authority and
responsibility for schedule, cost, safety, and technical requirements and that
this was not an effective check and balance.[21] The CAIB report
recommended that NASA establish a technical authority to serve
independently of the Space Shuttle program so that employees would not
feel hampered to bring forward safety concerns or disagreements with
programmatic decisions. The Board's findings that led to this
recommendation included a broken safety culture in which it was difficult
for minority and dissenting opinions to percolate up through the hierarchy;
dual Center and programmatic roles vested in one person that had confused
lines of authority, responsibility, and accountability and made the oversight
process susceptible to conflicts of interest; and oversight personnel in
positions within the program, increasing the risk that these staffs'
perspectives would be hindered by too much familiarity with the programs
they were overseeing.

ESD officials stated that they had carefully and thoughtfully
implemented the intent of the CAIB; they said they had not disregarded its
finding and recommendations but instead established a technical authority

---

[21] *Columbia Accident Investigation Board Report*, Volume I (Washington, D.C.: August 2003).The
CAIB report also addressed the findings of the Rogers Commission, which was created after
the Challenger accident in 1986 to investigate the cause of the accident. The CAIB reported
that the Rogers Commission's findings identified cost and schedule pressures and the lack of
independent safety oversight at NASA as contributing factors to the Challenger accident.

in such a way that it best fit the context of ESD's efforts. These officials did acknowledge, though, that the dual hat approach does not align with the CAIB report's recommendation to separate programmatic and technical authority or with NASA's governance framework. Further, over the course of our review, we spoke with various high-ranking officials outside and within NASA who expressed some reservations about ESD's dual hat approach. For example:

- The former Chairman of the CAIB stated that, even though the ESD programs are still in development, he believes the technical authority should be institutionally protected against the pressures of cost and schedule and added that NASA should never be lulled into dispensing of engineering and safety independence because human spaceflight is an extremely risky enterprise.
- Both NASA's Chief Engineer and Chief of S&MA acknowledged there is inherent conflict in the concurrent roles of the dual hats, while also expressing great confidence in the ESD staff now in the dual roles.
- NASA's Chief of S&MA indicated that the dual-hat S&MA structure is working well within ESD, but he believes these dual-hatted roles may not necessarily meet the intent of the CAIB's recommendation because the Board envisioned an independent safety organization completely outside the programs.
- NASA's Chief Engineer stated that he believes technical authority should become a separate responsibility and position as ESD moves forward with integration of the three programs and into their operation as a system.

As these individuals made clear, ensuring the ESD engineering and S&MA technical authorities remain independent of cost and schedule conflicts is key to human spaceflight success and safety. Along these lines, the ASAP previously conveyed concerns about NASA's implementation of technical authority that continue to be valid today. In particular, the ASAP stated in a 2013 report that NASA's technical authority was working at that time in large measure due to the well-qualified, strong personnel that had been assigned to the process.[22] The panel noted, however, that should there be a conflict or weakening of the placement of strong individuals in the technical authority position, this could introduce greater risk into a program. Although a current ASAP official stated she had no concerns with ESD's present approach to technical authority, the panel's prior caution remains applicable, and the risk that the ASAP identified earlier could be realized if not mitigated by eliminating the potential for competing interests within the ESD engineering and S&MA positions.

NASA is currently concluding an assessment of the implementation of the technical authority role to determine how well that function is working across the agency. According to the official responsible for leading the study, the assessment includes examining the evolution of the technical authority role over the years and whether NASA is spending the right amount of funds for those positions. NASA expects to have recommendations in 2017 on how to improve the technical authority function, but does not expect to address the dual hat construct. A principle of federal internal controls is that an agency should design control activities to achieve objectives and respond to risks, which includes segregation of key duties and responsibilities to reduce the risk of error, misuse, or fraud.[23] By overlapping technical authority and programmatic responsibilities, NASA will continue to run the risk of creating an environment of competing interests for the ESD engineering and S&MA technical authorities.

ESD RISK POSTURE HAS IMPROVED, BUT KEY RISK AREAS REMAIN FOR THE INTEGRATION EFFORT

Despite the development and integration challenges associated with a new human spaceflight capability, ESD has improved its overall cross-

---

[22] NASA, Aerospace Safety Advisory Panel, *Annual Report for 2012* (Washington, D.C.: Jan. 9, 2013).
[23] GAO, *Standards for Internal Control in the Federal Government,* GAO-14-704G (Washington, D.C.: September 2014).

program risk posture over the past 2 years. Nonetheless, it still faces key integration risk areas within software development and verification and validation (V&V). Both are critical to readiness for EM-1 because software acts as the "brain" that ties SLS, Orion, and EGS together in a functioning body, while V&V ensures the integrated body works as expected. The success of these efforts forms the foundation for a launch, no matter the date of EM-1.

ESD's Cross-Program Risk Posture Has Improved

We have previously reported on individual SLS, Orion, and EGS program risks that were contributing to potential delays within each program.[24] For example, in July 2016, we found that delays with the European Service Module—which will provide services to the Orion crew module in the form of propulsion, consumables storage, and heat rejection and power—could potentially affect the Orion program's schedule.[25] Subsequently, in April 2017, we found that those delays had worsened and were contributing to the program likely not making a November 2018 launch readiness date.[26] All three programs continue to manage such individual program risks, which is to be expected of programs of this size and complexity. The programs may choose to retain these risks in their own risk databases or elevate them to ESD to track mitigation steps. A program would elevate a risk to ESD when decisions are needed by ESD management, such as a need for additional resources or requirement changes. Risks with the greatest potential for negative impacts are categorized as top ESD risks. In addition to these individual programs risks that are elevated to ESD, ESD is also responsible for overseeing cross-program risks that affect multiple programs. An example of a cross-program risk is the potential for delayed

---

[24] GAO-17-414; GAO-16-612; and *Orion Multi-Purpose Crew Vehicle: Action Needed to Improve Visibility into Cost, Schedule, and Capacity to Resolve Technical Challenges*, GAO-16-620 (Washington, D.C.: July 27, 2016).

[25] GAO-16-620.

[26] GAO-17-414.

delivery of data from SLS and Orion to affect the EGS software development schedule.

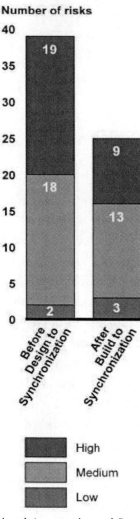

Source: GAO analysis of National Aeronautics and Space Administration data. | GAO-18-28.

Figure 10. Exploration Systems Development Organization's Progress in Reducing Risks, 2014-2017.

ESD has made progress reducing risks over the last 2 years, from the point of the Design to Sync preliminary design review equivalent for the integrated programs to the Build to Sync critical design review equivalent. As figure 10 illustrates, ESD has reduced its combined total of ESD and cross program risks from 39 to 25 over this period, and reduced the number of high risks from about 49 percent of the total to about 36 percent of the total.[27]

The ESD risk system is dynamic, with risks coming into and dropping out of the system over time as development proceeds and risk mitigation is completed. A total of 29 of the 39 risks within the ESD risk portfolio were removed from the register and 15 risks were added to the register between November 2014, prior to Design to Sync, and March 2017, after Build to Sync. Examples of risks removed over this time period include risks associated with late delivery of Orion and SLS ground support equipment hardware to EGS and establishing a management process to identify risks stemming from the programs being at differing points in development.

Nine risks remained active in the system over the 2-year period we analyzed, and NASA experienced delays in the length of time it anticipated it would take to complete mitigation of the majority of these nine risks. Three of these nine risks that have remained active in the risk system since before Design to Sync are still classified as high risk; the remaining six are classified as medium risk. Mitigation is an action taken to eliminate or reduce the potential severity of a risk, either by reducing the probability of it occurring, by reducing the level of impact if it does occur, or both. ESD officials indicated a number of reasons why risks could take longer to mitigate. For instance, risks with long-term mitigation strategies may go for extended periods of time without score changes. In addition, ESD may conduct additional risk assessments and determine that certain risks need to be reprioritized over time and that resources should be focused towards higher risks. In addition, some risk mitigation steps are tied to hardware delivery and launch dates, and as those delay, the risk mitigation steps will

---

[27] Risks categorized as high have the greatest potential for major impacts to cost, schedule, performance or safety. Medium risks have the potential for moderate impacts and low risks have the potential for minor impacts.

as well. As illustrated in table 2, we found that six of these nine risks were related to software and V&V and represented some of the primary causes in terms of estimated completion delays. On average, the estimated completion dates for these six risks were delayed about 16 months. In addition, the two V&V risks that have remained active since before Design to Sync were still considered top ESD risks as of March 2017 when we completed this analysis.

Table 2. Change in Estimated Completion Date for Nine Exploration Systems Development Organization's Risks Active from before Design to Sync to after Build to Sync

Risk	Risk description	Changes in status of risk	Changes in estimated completion date from Design to Sync to Build to Sync and NASA rationale for changes
Application Software for Multi Payload Processing Facility	Software. Delays in delivery of cross-program products to the Exploration Ground Systems (EGS) program, including hardware and models from the Space Launch System (SLS) and Orion programs, could increase the likelihood that the ground software may not be ready to support Exploration Mission1 (EM-1) processing.	Decreasing	24 months Mitigation plan and schedule changes are a result of new technical information surfacing, additional work required, and software schedule re-planning.
Data Throughput	Software. Spaceport Command and Control System—a ground software system that controls ground equipment; records and retrieves data from systems before and during launch; and monitors the health and status of spacecraft as they prepare for and launch—may not be able to process the amount of instrument readings received and provide commands to SLS and ground equipment as required.	Increasing	22months Mitigation plan and schedule changes are a result of new technical information surfacing, additional work required, and software schedule re-planning.

Risk	Risk description	Changes in status of risk	Changes in estimated completion date from Design to Sync to Build to Sync and NASA rationale for changes
Cryogenic Operations Application Software Development	Software. Final testing of the SLS core stage before it is shipped to Kennedy Space Center is likely to identify the need for changes to the ground software controlling cryogenic operations—which could be so substantial that EGS has insufficient resources and time in the schedule allocated to meet the EM-1 launch schedule.	No Change	14 months New technical information surfaced and delay in delivery schedules from other programs resulted in mitigation plan updates.
Application Software for Mobile Launcher/ Vehicle Assembly Building Integrated Processing	Software. EGS may not receive requirements and data products required to finalize Mobile Launcher and Vehicle Assembly Building software from SLS and Orion in time to support the launch date.	No Change	Estimated closure date moved up approximately1 month Mitigation plan and schedule changes are a result of new technical information surfacing, additional work required, and software schedule re-planning.
Insufficient Schedule for Verification and Validation at Kennedy Space Center	Verification and Validation. There are significant threats to the verification and validation schedule and budget due to schedule and cost baselines not accounting for rework, redesign, testing, and problem resolution.	Decreasing	19 months Mitigation plan schedule is coupled with the program baseline schedule; therefore, any shift in baseline schedule moves out mitigation plan schedule.
Space Launch System Integrated Loads Modeling May Delay Launch Date	Verification and Validation. The SLS program is not conducting integrated dynamics load testing. Instead it is testing components separately and developing models based on this testing. The Exploration Systems Development (ESD) organization is conducting integrated dynamics testing on the flight hardware at Kennedy Space Center after stacking.	No Change	12 months Most of the risk mitigation steps are paced by testing activities, which have been delayed due to hardware manufacturing difficulties. The mitigation plan is coupled to the schedule and will move as the program re-baselines.

Table 2. (Continued)

Risk	Risk description	Changes in status of risk	Changes in estimated completion date from Design to Sync to Build to Sync and NASA rationale for changes
	If problems with the SLS dynamic loads models are discovered at Kennedy Space Center, EM-1 launch may be delayed.		
Integrated Operations for EM-1	Operations. There may be a learning curve associated with launching a new integrated system for the first time.	Decreasing	5 months Mitigation plan schedule is coupled with the Program Baseline schedule; therefore, any shift in baseline schedule moves out mitigation plan schedule.
Launch Abort Vehicle Limitations	Hardware. Orion's launch abort system may not function during all phases of launch on the SLS.	No Change	Estimated closure date moved less than one month This risk is being held open until a dedicated ascent abort test is completed per the original mitigation plan.
SLS Booster propellant liner Insulation Structures/Fracture Behavior	Hardware. New propellant, liner, and insulation materials may be unable to satisfy agency requirements for structural/fracture certification.	No Change	Completion date not established prior to Design to Sync review. New technical information surfaced such that additional work was required. Complex mitigation roadmap established that is currently being worked to demonstrate acceptable risk.

Design to Sync= Design to Synchronization
Build to Sync= Build to Synchronization
Source: GAO analysis of National Aeronautics and Space Administration data. | GAO-18-28.

Software Development Is a Key Risk Area Facing the Integration Effort

Software development is one of the top cross-program technical issues facing ESD as the programs approach EM-1. Software is a key enabling technology required to tie the human spaceflight systems together. Specifically, for ESD to achieve EM-1 launch readiness, software developed within each of the programs has to be able to link and communicate with software developed in other programs in order to enable a successful launch. Furthermore, software development continues after hardware development and is often used to help resolve hardware deficiencies discovered during systems integration and test.

ESD has defined six critical paths—the path of longest duration through the sequence of activities that determines the program's earliest completion date—for its programs to reach EM-1, and three are related to software development. These three software critical paths support interaction and communication between the systems the individual programs are developing—SLS to EGS software, Orion to EGS software, and the Integrated Test Laboratory (ITL) facility that supports Orion software and avionics testing as well as some SLS and EGS testing. The other critical paths are development of the Orion crew service module, SLS core stage, and the EGS Mobile Launcher. Because of software's importance to EM-1 launch readiness, ESD is putting a new method in place to measure how well these software efforts are progressing along their respective critical paths. To that end, it is currently developing a set of "Key Progress Indicators" milestones that will include baseline and forecast dates. Officials indicated that these metrics will allow ESD to better track progress of the critical path software efforts toward EM-1 during the remainder of the system integration and test phase. ESD officials have indicated, however, that identifying and establishing appropriate indicators is taking longer than expected and proving more difficult than anticipated.

One of the software testing critical paths, the ITL, has already experienced delays that slipped completion of planned software testing from September 2018 until March 2019, a delay of 6 months. Officials told us that this delay was primarily due to a series of late avionics and software deliveries by the European Space Agency for Orion's European Service Module. The delay in the Orion testing in turn affects SLS and EGS software testing and integration because those activities are informed by the completion of the Orion software testing. Furthermore, some EGS and SLS software testing scheduled to be conducted within the ITL has been re-planned as a result of the Orion delays.

The Orion program indicates that it has taken action to mitigate ITL issues as they arise. For example, the European Service Module avionics and software delivery delay opened a 125-day gap between completion of crew module testing and service module testing. Orion officials indicated that the program had planned to proceed directly into testing of the integrated crew module and service module software and systems, but the integrated testing cannot be conducted until the service module testing is complete.

As illustrated by figure 11, to mitigate the impact of the delay, Orion officials indicated that the program filled this gap by rescheduling other activities at the ITL such as software integration testing and dry runs for the three programs.[28] These adjustments narrowed the ITL schedule gap from 125 days to 24 days. The officials stated that they will continue to adjust the schedule to eliminate gaps.

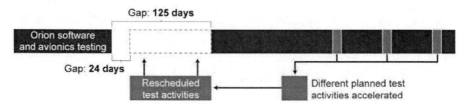

Source: GAO analysis of National Aeronautics and Space Administration data. | GAO-18-28.

Figure 11. Orion Software and Avionics Testing at Integrated Testing Lab.

The other two software critical paths—SLS to EGS and Orion to EGS software development—are also experiencing software development issues. In July 2016, for example, we found that delays in SLS and Orion

---

[28] NASA officials indicated that testing dry runs are conducted to ensure the test setup and procedures are mature enough to proceed into formal test events.

requirements development, as well as the programs' decisions to defer software content to later in development, were delaying EGS's efforts to develop ground command and control software and increasing cost and schedule.[29]

Furthermore, ESD reports show that delays and content deferral in the Orion and SLS software development continue to affect EGS software development and could delay launch readiness. For example, the EGS data throughput risk that both ESD and EGS are tracking is that the ground control system software is currently not designed to process the amount of telemetry it will receive and provide commands to SLS and ground equipment as required during launch operations. EGS officials stated that, if not addressed, the risk is that if there is a SLS or Orion failure, the ground control system software may not display the necessary data to launch operations technicians. EGS officials told us that the reason for the mismatch between the data throughput being sent to the ground control software and how much is it designed to process is that no program was constrained in identifying its data throughput. These officials stated that retrospectively, they should have established an interface control document to manage the process. The officials also stated that the program is taking steps to mitigate this risk, including defining or constraining the data parameters and buying more hardware to increase the amount of data throughput that can be managed, but will not know if the risk is fully mitigated until additional data are received and analyzed during upcoming tests. For example, EGS officials stated that the green run test will provide additional data to help determine if the steps they are taking address this throughput risk.[30] If the program determines the risk is not fully mitigated and additional software redesign is required, it could lead to schedule delays.

ESD officials overseeing software development acknowledged that software development for the integrated systems is a difficult task and said they expect to continue to encounter and resolve software development

---

[29] GAO-16-612.

[30] Green run is the culminating test of the SLS core stage development where the actual EM-1 core stage flight article will be integrated with the cluster of four RS-25 engines and fired for 500 seconds under simulated flight conditions.

issues during cross-program integration and testing. As we have found in past reviews of NASA and Department of Defense systems, software development is a key risk area during system integration and testing. For example, we found in April 2017 that software delivery delays and development problems with the U.S. Air Force's F-35 program experienced during system integration and testing were likely to extend that program's development by 12 months and increase its costs by more than $1.7 billion.[31]

Verification and Validation Will Remain Key Risk Area to Monitor as NASA Establishes and Works towards New Launch Readiness Date

Verification and validation (V&V) is acknowledged by ESD as a top cross-program integration risk that NASA must monitor as it establishes and works toward a new EM-1 launch readiness date. V&V is a culminating development activity prior to launch for determining whether integrated hardware and software will perform as expected. V&V consists of two equally important aspects:

- verification is the process for determining whether or not a product fulfills the requirements or specifications established for it at the start of the development phase; and
- validation is the assessment of a planned or delivered system ability to meet the sponsor's operational need in the most realistic environment achievable during the course of development or at the end of development.

Like software development and testing, V&V is typically complex and made even more so by the need to verify and validate how SLS, Orion, and EGS work together as an integrated system.

---

[31] GAO- F-35 *JOINT STRIKE FIGHTER: DOD Needs to Complete Developmental Testing before Making Significant New Investments,* GAO-17-351 (Washington, D.C.: Apr. 24, 2017).

ESD's V&V plans for the integrated system have been slow to mature. In March 2016, leading up to ESD's Build to Sync review, ESD performed an audit of V&V-related documentation for the program CDRs and ESD Build to Sync. The audit found that 54 of 257 auditable areas (21 percent) were not mature enough to meet NASA engineering policy guidance for that point in development. According to ESD documentation, there were several causes of this immaturity, including incomplete documentation and inconsistent requirements across the three programs. NASA officials told us that our review prompted ESD to conduct a follow-up and track the status of these areas. As of June 2017, 53 of the 54 auditable areas were closed, which means these areas are at or have exceeded CDR level of maturity—6 months after Build to Sync was completed. NASA officials indicated that the remaining one auditable area, which is related to the test plan for the integrated communication network, was closed in August 2017.

Nevertheless, other potential V&V issues still remain. According to ESD officials, distributing responsibility for V&V across the three programs has created an increased potential for gaps in testing. If gaps are discovered during testing, or if integrated systems do not perform as planned, money and time for modifications to hardware and/or software may be necessary, as well as time for retesting. This could result in delayed launch readiness. As a result, mature V&V plans are needed to ensure there are no gaps in planned testing. ESD officials indicated that a NASA Engineering and Safety Center review of their V&V plans, requested by ESD's Chief Engineer to address concerns about V&V planning, would help define the path forward for maturing V&V plans. V&V issues add to cost and schedule risk for the program because they may take more time and money to resolve than ESD anticipates. In some cases, they may have a safety impact as well. For example, if the structural models are not sufficiently verified, it increases flight safety risks. Each of the programs bases its individual analyses on the models of the other programs. As a result, any deficiencies discovered in one can have cascading effects through the other systems and programs. We will continue to monitor ESD's progress mitigating risks as NASA approaches EM-1.

CONCLUSION

NASA is at the beginning of the path leading to human exploration of Mars. The first phase along that path, the integration of SLS, Orion, and EGS, is likely to set the stage for the success or failure of the rest of the endeavor. Establishing a cost and schedule baseline for NASA's second mission is an important initial step in understanding and gaining support for the costs of SLS, Orion, and EGS, not just for that one mission but for the Mars plan overall. NASA's ongoing refusal to establish this baseline is short-sighted, because EM-2 is part of a larger conversation about the affordability of a crewed mission to Mars. While later stages of the Mars mission are well in the future, getting to that point in time will require a funding commitment from the Congress and other stakeholders. Much of their willingness to make that commitment is likely to be based on the ability to assess the extent to which NASA has met prior goals within predicted cost and schedule targets.

Furthermore, as ESD moves SLS, Orion, and EGS from development to integrated operations, its efforts will reach the point when human lives will be placed at risk. Space is a severe and unforgiving environment; the Columbia accident showed the disastrous consequences of mistakes. As the Columbia Accident Investigation Board report made clear, a program's management approach is an integral part of ensuring that human spaceflight is as safe and successful as possible. The report also characterized independence as key to achieving that safety and success. ESD's approach, however, tethers independent oversight to program management by vesting key individuals to wear both hats at the same time. As a result, NASA is relying heavily on the personality and capability of those individuals to maintain independence rather than on an institutional process, which diminishes lessons learned from the Columbia accident.

MATTER FOR CONGRESSIONAL CONSIDERATION

We are making the following matter for congressional consideration.

Congress should consider requiring the NASA Administrator to direct the Exploration Systems Development organization within the Human Exploration and Operations Mission Directorate to establish separate cost and schedule baselines for work required to support SLS and EGS for Exploration Mission 2 and establish separate cost and schedule baselines for each additional capability that encompass all life cycle costs, to include operations and sustainment. (Matter for Consideration 1)

RECOMMENDATION FOR EXECUTIVE ACTION

We are making the following recommendation to the Exploration Systems Development organization.

Exploration Systems Development should no longer dual-hat individuals with both programmatic and technical authority responsibilities. Specifically, the technical authority structure within Exploration Systems Development should be restructured to ensure that technical authorities for the Offices of the Chief Engineer and Safety and Mission Assurance are not fettered with programmatic responsibilities that create an environment of competing interests that may impair their independence. (Recommendation 1)

AGENCY COMMENTS AND OUR EVALUATION

NASA provided written comments on a draft of this chapter. These comments are reprinted in appendix II. NASA also provided technical comments, which were incorporated as appropriate.

In responding to a draft of our report, NASA partially concurred with our recommendation that the Exploration Systems Development (ESD) organization should no longer dual-hat individuals with both programmatic and technical authority responsibilities. Specifically, we recommended that the technical authority structure within ESD should be restructured to ensure that technical authorities for the Offices of Chief Engineer and Safety and Mission Assurance are not fettered with programmatic responsibilities that create an environment of competing interests that may impair their independence. In response to this recommendation, NASA stated that it created the technical authority governance structure after the Columbia Accident Investigation Board report and that the dual-hat technical authority structure has been understood and successfully implemented within ESD. NASA recognized, however, that as the program moves from the design and development phase into the integration and test phase, it anticipates that the ESD environment will encounter more technical issues that will, by necessity, need to be quickly evaluated and resolved. NASA asserted that within this changed environment it would be beneficial for the Engineering Technical Authority role to be performed by the Human Exploration and Operations Chief Engineer (who reports to the Office of the Chief Engineer). NASA stated that over the next year or so, it would solicit detailed input from these organizations and determine how to best support the program while managing the transition to integration and test and anticipated closing this recommendation by September 30, 2018.

We agree that NASA should solicit detailed input from key organizations within the agency as it transitions away from the dual hat technical authority structure to help ensure successful implementation of a new structure. In order to implement this recommendation, however, NASA needs to assign the technical authority role to a person who does not have programmatic responsibilities to ensure they are independent of responsibilities related to cost and schedule performance. To fulfill this, this person may need to reside outside of the Human Exploration and Operations Mission Directorate and NASA should solicit input from the Office of the Chief Engineer when making this decision to ensure that there are no competing interests for the technical authority. Moreover, in its response, NASA does not address the dual-hat technical authority role for Safety and

Mission Assurance. We continue to believe that similar changes for this role would be appropriate as well.

Further, in response to this recommendation, NASA makes two statements that require additional context. First, NASA stated that GAO's recommendation was focused on overall Agency technical authority management. While this review involved meeting with the heads of the Office of Chief Engineer and the Office of Safety and Mission Assurance, the scope of this review and the associated recommendation are limited to ESD. Second, NASA stated "As you found, we agree that having the right personnel in senior leadership positions is essential for a Technical Authority to be successful regardless of how the Technical Authority is implemented." To clarify, this perspective is attributed to NASA officials in our report and does not represent GAO's position.

Cristina T. Chaplain
Director, Acquisition and Sourcing Management

APPENDIX I: OBJECTIVES, SCOPE, AND METHODOLOGY

This chapter assesses (1) the benefits and challenges of the National Aeronautics and Space Administration's (NASA) approach for integrating and assessing the programmatic and technical readiness of Orion, SLS, and EGS; and (2) the extent to which the Exploration Systems Development (ESD) organization is managing cross-program risks that could affect launch readiness.

To assess the benefits and challenges of NASA's approach for integrating and assessing the programmatic and technical readiness of its current human spaceflight programs relative to other selected programs, we reviewed and analyzed NASA policies governing program and technical integration, including cost, schedule, and risk. We obtained and analyzed ESD implementation plans to assess the role of ESD in cross program integration of the three programs. We reviewed the 2003 Columbia Accident

Investigation Board's Report's findings and recommendations related to culture and organizational management of human spaceflight programs as well as the Constellation program's lessons learned report. We reviewed detailed briefings and documentation from Cross-Program Systems Integration and Programmatic and Strategic Integration teams explaining ESD's approach to programmatic and technical integration, including implementation of systems engineering and integration. We interviewed NASA officials to discuss the benefits and challenges of NASA's integration approach and their roles and responsibilities in managing and overseeing the integration process. We met with the technical authorities and other representatives from the NASA Office of the Chief Engineer, Office of Safety and Mission Assurance, Crew Health and Safety, addressed cost and budgeting issues with the Chief Financial Officer and discussed and documented their roles in executing and overseeing the ESD programs. We also interviewed outside subject matter experts to gain their insight of ESD's implementation of NASA's program management policies on the independent technical authority structure. Additionally, we compared historical budget data from the now-cancelled Constellation program to ESD budget data and quantified systems engineering and integration budget savings through preliminary design review, the point at which the Constellation program was cancelled. In addition, we assessed the scope of NASA's funding estimates for the second exploration mission and beyond against best practices criteria outlined in GAO's cost estimating guidebook.[32] We assessed the reliability of the budget data obtained using GAO reliability standards as appropriate. We compared the benefits and challenges of NASA's integration approach to that of other complex, large-scale government programs, including NASA's Constellation and the Department of Defense's Missile Defense Agency programs.

To determine the extent to which ESD is managing cross-program risks that could affect launch readiness, we obtained and reviewed NASA and ESD risk management policies; detailed monthly and quarterly briefings;

---

[32] GAO-09-3SP.

and documentation from Cross-Program Systems Integration and Programmatic and Strategic Integration teams explaining ESD's approach to identifying, tracking, and mitigating cross-program risks. We reviewed Cross-Program Systems Integration systems engineering and systems integration areas as well as Programmatic and Strategic Integration risks, cost, and schedule to determine what efforts presented the highest risk to cross program cost and schedule. We conducted an analysis of ESD's risk dataset and the programs' detailed risk reports, which list program risks and their potential schedule impacts, including mitigation efforts to date. We examined risk report data from Design to Sync to Build to Sync and focused our analyses to identify risks with current mitigation plans to determine if risk mitigation plans are proceeding on schedule. We did not analyze risks that were categorized under "Accept," "Candidate," "Research," "Unknown," or "Watch" because these risks were not assigned an active mitigation plan by ESD. To assess the reliability of the data, we reviewed related documentation and interviewed knowledgeable agency officials. We determined the data was sufficiently reliable for identifying risks and schedule delays associated with those risks. We examined ESD integrated testing facility schedules to determine the extent to which they can accommodate deviation in ESD's planned integrated test schedule. We also interviewed program and contractor officials on technical risks, potential impacts, and risk mitigation efforts underway and planned.

We conducted this performance audit from August 2016 to October 2017 in accordance with generally accepted government auditing standards. Those standards require that we plan and perform the audit to obtain sufficient, appropriate evidence to provide a reasonable basis for our findings and conclusions based on our audit objectives. We believe that the evidence obtained provides a reasonable basis for our findings and conclusions based on our audit objectives.

APPENDIX II: COMMENTS FROM THE NATIONAL AERONAUTICS AND SPACE ADMINISTRATION

National Aeronautics and Space Administration

Headquarters
Washington, DC 20546-0001

OCT -6 2017

Reply to Attn of: Human Exploration and Operations Mission Directorate

Ms. Cristina T. Chaplain
Director
Acquisition and Sourcing Management
United States Government Accountability Office
Washington, DC 20548

Dear Ms. Chaplain:

The National Aeronautics and Space Administration (NASA) appreciates the opportunity to review and comment on the Government Accountability Office (GAO) draft report entitled, "NASA Human Space Exploration: Integration Approach Presents Challenges to Oversight and Independence" (GAO-18-28), dated August 22, 2017.

Progress made in Exploration Systems Development (ESD) Programs represents a significant achievement for the Agency and the nation's future. The systems development, fabrication, and assembly work being performed today is setting the basis for a series of missions that will lead to the Moon, Mars, and beyond. We recognize that there are inherent technical risks to any endeavor of such significance. GAO has noted the strengths of the NASA risk management system, and NASA will continue to be vigilant in our efforts to manage risk.

NASA recognizes the value of GAO's objective evaluation of the ESD approach to program integration, and we are encouraged by the findings concerning the cost savings and efficiency of the ESD integrated management approach. We are also encouraged that, following this most recent, extensive programmatic and technical review, GAO's recommendation was focused on overall Agency technical authority management.

In the report, GAO makes the following recommendation to NASA, along with the following Matter for Congressional Consideration:

Recommendation: ESD should no longer dual-hat individuals with both programmatic and technical authority responsibilities. Specifically, the technical authority structure within ESD should be restructured to ensure that technical authorities for the Offices of the Chief Engineer and Safety and Mission Assurance are not fettered with programmatic responsibilities that create an environment of competing interests that may impair their independence.

Management's Response: NASA partially concurs with this recommendation. NASA appreciates the GAO's thorough review of the Agency Technical Authority (TA) process as part of the ESD integration audit. As you found, we agree that having the right personnel in senior leadership positions is essential for a TA to be successful regardless of how TA is implemented. After the Columbia Accident Investigation Board (CAIB) report was issued, NASA created the NASA Engineering and Safety Center (NESC), the Technical Authority (TA) governance structure, and the NASA Safety Center to assure the highest caliber candidates in TA positions. The ESD TA positions have been filled by Senior Executive Service (SES) and Senior Level/Scientific or Professional (ST/SL) civil service personnel who retained independence because their reporting paths and performance reviews are all within the respective TA chain of command, a command that is also led by highly qualified, career SES experts in their designated field.

The dual-hat TA structure has been understood and successfully implemented since the inception of the ESD Enterprise. Any changes to this structure must be carefully considered with substantial input from Center Directors, the Acting Administrator, the Programs, the TAs, and NASA's Human Capital Management organizations. As the ESD Enterprise moves from the design/development phase of the program to the integration and test phase, we anticipate an environment in which more technical issues arise that will, by necessity, need to be quickly evaluated and resolved. As the Programs adjust to this increasing technical workload, it would be beneficial for the Director for Cross-Program Systems Integration (CSI) to adapt and maintain the programmatic Systems Engineering and Integration (SE&I) lead for the Enterprise and have the Engineering TA role performed by the Human Exploration and Operations Chief Engineer (who reports to the Office of the Chief Engineer). In the next year or so, NASA will solicit detailed input from these organizations and determine how to best support the program while managing the transition to integration and test.

Estimated Completion Date: September 30, 2018

In addition to this programmatic recommendation, GAO also proposed a Matter for Congressional Consideration. GAO has raised this issue of changing the basis for the Agency's previously-approved exploration commitment framework. NASA's position on this matter has not changed and remains consistent with both Agency policy and direction in prior authorization acts.

Matter for Congressional Consideration: Congress should consider requiring the NASA Administrator to direct the Exploration systems development organization within the Human Exploration and Operations Mission Directorate to establish separate cost and schedule baselines for work required to support Space Launch System (SLS) and Exploration Ground Systems (EGS) for Exploration Mission-2 (EM-2) and establish separate cost and schedule baselines for each additional capability that encompass all life cycle costs, to include operations and sustainment (Matter for Consideration 1).

Management's Response: It is critical for Agency officials and external stakeholders to understand the nature and goals of the ESD enterprise, its progress towards mission objectives, and the costs involved. ESD is a long-term, multi-mission program to establish the space exploration infrastructure required to meet national goals. It is both a major new human spaceflight development and a capability that will evolve over time. This requires management approaches and performance metrics that are different from a one-off single mission project. NASA has created separate programs for SLS, Orion, and Ground Systems and Development Operations (GSDO), and each program's development is paced to when a particular capability is needed, consistent with the direction from the Administration and Congress. For example, as workforce completes pieces comprising the first flight of SLS or Orion, they move on to those needed for the next flight. They are not held up by a different pace on the other programs. NASA has also adopted a block upgrade approach for SLS to ensure realistic long-range investment planning and effective resource allocations through the budget process. NASA regularly balances available funding with the flight manifest within the context of the Agency's overall exploration objectives. NASA's programmatic decisions are based on optimizing acquisition strategies and resource allocations (material, people, funding) across multiple missions to ensure efficient implementation of deep space exploration objectives that take several flights to accomplish. NASA believes it has the processes in place to provide stakeholders insight to cost, schedule, and risks that accord with ESD's nature as a multi-mission space transportation infrastructure. Cost estimates and expenditures are available for future missions; however, these costs must be derived from the data and are not directly available. This was done by design to lower NASA's expenditures. NASA does not think that structuring acquisition and implementation to ease accounting on a mission-by-mission basis is prudent as it would result in higher overall program costs and is not in keeping with the nature of the program.

Once again, thank you for the opportunity to review and comment on the subject draft report. If you have any questions or require additional information, please contact Lynne Loewy on (202) 358-0549.

Sincerely,

William H. Gerstenmaier
Associate Administrator
 for Human Exploration and Operations

APPENDIX III: STAFF ACKNOWLEDGMENTS

In addition to Cristina T. Chaplain, Molly Traci (Assistant Director), LaTonya Miller, John S. Warren Jr., Tana Davis, Laura Greifner, Roxanna T. Sun, Samuel Woo, Marie P. Ahearn, and Lorraine Ettaro made key contributions to this chapter.

APPENDIX IV: ACCESSIBLE DATA

Data Tables

Accessible Data for Competing Interests between Engineering Technical Authority Role and Program Role

Technical authority	Program authority
Engineering	Engineering
Acts on behalf of the NASA Office of the Chief Engineer to provide independent oversight and assessment of ESD engineering-related matters	As ESD Director of Cross Program System Integration, acts on behalf of ESD by owning and managing ESD's integrated technical baseline and managing ESD's resources to accomplish it

ESD = Exploration Systems Development
NASA = National Aeronautics and Space Administration

Accessible Data for Figure 2: Select Components of Exploration Ground Systems Program

•	Mobile Launcher
•	Launch Pad 39B
•	Vehicle Assembly Building
•	Vehicle Assembly Building, inside High Bay 3
•	Crawler-Transporter

Accessible Data for Figure 3: NASA's Life Cycle for Space Flight Projects

Formulation			Implementation			
KDP A	KDP B	KDP C (confirmation review) Project start	KDP D	KDP E	KDP F	n/a
Pre-phase A Concept studies	Phase A Concept and technology development	Phase B Preliminary design and technology completion	Phase C Final design and fabrication	Phase D System assembly, integration and test, and launch	Phase E Operations and sustainment	Phase F Closeout

Accessible Data for Figure 3. (Continued)

Formulation			Implementation			
n/a	SDR/MDR	PDR	CDR	SIR	n/a	n/a
Management decision reviews • KDP = key decision point Technical reviews • SDR/MDR = system definition review/mission definition review • PDR = preliminary design review • CDR = critical design review • SIR = system integration review						

Accessible Data for Figure 6: Exploration Systems Development Organization's Integration Reviews

Formulation			Implementation			
Pre-phase A Concept studies	Phase A Concept and technology development	Phase B Preliminary design and technology completion	Phase C Final design and fabrication	Phase D System assembly, integration and test, and launch	Phase E Operations and sustainment	Phase F Closeout
Cross-program SRR	ESI system definition review	PDR ESI design-to-sync	CDR ESI build-to-sync	EM-1 Mission integration review	EM-1 Integration review	n/a
• SRR = System requirements review • PDR = Preliminary design review • CDR = Critical design review • ESI = Exploration systems integration • EM-1 = Exploration Mission-1						

Accessible Data for Figure 7: Exploration Systems Development Organization's Integration Budget Fiscal Years 2012-2017

Year	Estimate (in millions)	Actuals (in millions)
2012	29.6	16.1
2013	29.6	28.6
2014	29.6	30.5
2015	29.6	32
2016	29.6	33.6
2017	29.6	38.9

Accessible Data for Figure 8: Exploration Systems Development Organization's Configuration Management Records Approval Rate

Year	Multiple Month Time Period	Original Planned approval Rate	New Planned approval rate	Actuals
2016	Time Period Mar -June	approval Rate218	approval rate218	208
	July-Sept	298	298	157
	Oct-Dec	181	181	138
2017	Jan -Mar	277	197	138
	Apr -June	351	230	281
	July-Sept	602	277	n/a
	Oct-Dec	510	331	n/a
2018	Jan -Mar	402	343	n/a
	Apr -June	104	321	n/a
	July-Sept	98	285	n/a
	Oct-Dec	0	240	n/a
2019	Jan -Mar	0	170	n/a
	Apr -June	0	98	n/a
	July-Sept	0	99	n/a

Accessible Data for Figure 10: Exploration Systems Development Organization's Progress in Reducing Risks, 2014-2017

Low	Medium		High
Before Design to Synchronization	2	18	19
After Build to Synchronization	3	13	9

AGENCY COMMENT LETTER

Accessible Text for Appendix II: Comments from the National Aeronautics and Space Administration

National Aeronautics and Space Administration
Headquarters
Washington, DC 20546-0001
OCT 6 2017

Reply to Attn of: Human Exploration and Operations Mission Directorate Ms. Cristina T. Chaplain

Director

Acquisition and Sourcing Management

United States Government Accountability Office

Washington, DC 20548

Dear Ms. Chaplain:

The National Aeronautics and Space Administration (NASA) appreciates the opportunity to review and comment on the Government Accountability Office (GAO) draft report entitled, "NASA Human Space Exploration: Integration Approach Presents Challenges to Oversight and Independence" (GAO-18-28), dated August 22, 2017.

Progress made in Exploration Systems Development (ESD) Programs represents a significant achievement for the Agency and the nation's future. The systems development, fabrication, and assembly work being performed today is setting the basis for a series of missions that will lead to the Moon, Mars, and beyond. We recognize that there are inherent technical risks to any endeavor of such significance. GAO has noted the strengths of the NASA risk management system, and NASA will continue to be vigilant in our efforts to manage risk.

NASA recognizes the value of GAO's objective evaluation of the ESD approach to program integration, and we are encouraged by the findings concerning the cost savings and efficiency of the ESD integrated management approach. We are also encouraged that, following this most recent, extensive programmatic and technical review, GAO's recommendation was focused on overall Agency technical authority management.

In the report, GAO makes the following recommendation to NASA, along with the following Matter for Congressional Consideration:

Recommendation: ESD should no longer dual-hat individuals with both programmatic and technical authority responsibilities. Specifically, the technical authority structure within ESD should be restructured to ensure that technical authorities for the Offices of the Chief Engineer and Safety and Mission Assurance are not fettered with programmatic responsibilities that create an environment of competing interests that may impair their independence.

Management's Response: NASA partially concurs with this recommendation. NASA appreciates the GAO's thorough review of the Agency Technical Authority (TA) process as part of the ESD integration audit. As you found, we agree that having the right personnel in senior leadership positions is essential for a TA to be successful regardless of how TA is implemented. After the Columbia Accident Investigation Board (CAIB) report was issued, NASA created the NASA Engineering and Safety Center (NESC), the Technical Authority (TA) governance structure, and the NASA Safety Center to assure the highest caliber candidates in TA positions. The ESD TA positions have been filled by Senior Executive Service (SES) and Senior Level/Scientific or Professional (ST/SL) civil service personnel who retained independence because their reporting paths and performance reviews are all within the respective TA chain of command, a command that is also led by highly qualified, career SES experts in their designated field.

The dual-hat TA structure has been understood and successfully implemented since the inception of the ESD Enterprise. Any changes to this structure must be carefully considered with substantial input from Center Directors, the Acting Administrator, the Programs, the TAs, and NASA's Human Capital Management organizations. As the ESD Enterprise moves from the design/development phase of the program to the integration and test phase, we anticipate an environment in which more technical issues arise that will, by necessity, need to be quickly evaluated and resolved. As the Programs adjust to this increasing technical workload, it would be beneficial for the Director for Cross-Program Systems Integration (CSI) to adapt and maintain the programmatic Systems Engineering and Integration (SE&I) lead for the Enterprise and have the Engineering TA role performed by the Human Exploration and Operations Chief Engineer (who reports to the Office of the Chief Engineer). In the next year or so, NASA will solicit detailed input from these organizations and determine how to best support the program while managing the transition to integration and test.

Estimated Completion Date: September 30, 2018

In addition to this programmatic recommendation, GAO also proposed a Matter for Congressional Consideration. GAO has raised this issue of changing the basis for the Agency's previously-approved exploration commitment framework. NASA's position on this matter has not changed and remains consistent with both Agency policy and direction in prior authorization acts.

Matter for Congressional Consideration: Congress should consider requiring the NASA Administrator to direct the Exploration systems development organization within the Human Exploration and Operations Mission Directorate to establish separate cost and schedule baselines for work required to support Space Launch System (SLS) and Exploration Ground Systems (EGS) for Exploration Mission-2 (EM-2) and establish separate cost and schedule baselines for each additional capability that encompass all life cycle costs, to include operations and sustainment (Matter for Consideration 1).

Management's Response: It is critical for Agency officials and external stakeholders to understand the nature and goals of the ESD enterprise, its progress towards mission objectives, and the costs involved. ESD is a long-term, multi-mission program to establish the space exploration infrastructure required to meet national goals. It is both a major new human spaceflight development and a capability that will evolve over time. This requires management approaches and performance metrics that are different from a one-off single mission project. NASA has created separate programs for SLS, Orion, and Ground Systems and Development Operations (GSDO), and each program's development is paced to when a particular capability is needed, consistent with the direction from the Administration and Congress. For example, as workforce completes pieces comprising the first flight of SLS or Orion, they move on to those needed for the next flight. They are not held up by a different pace on the other programs. NASA has also adopted a block upgrade approach for SLS to ensure realistic long-range investment planning and effective resource allocations through the budget process. NASA regularly balances available funding with the flight manifest within the context of the Agency's overall exploration objectives. NASA's programmatic decisions are based on optimizing acquisition strategies and resource allocations (material, people, funding) across multiple missions to ensure efficient implementation of deep space exploration objectives that take several flights to accomplish. NASA believes it has the processes in place to provide stakeholders insight to cost, schedule, and risks that accord with ESD's nature as a multi-mission space transportation infrastructure. Cost estimates and expenditures are available for future missions; however, these costs must be derived from the data and are not directly available. This was done by design to lower NASA's expenditures. NASA does not think that structuring acquisition and implementation to ease accounting on a mission-by-mission basis is prudent as it would result in higher overall program costs and is not in keeping with the nature of the program.

Once again, thank you for the opportunity to review and comment on the subject draft report.

William H. Gerstenmaier
Associate Administrator for Human Exploration and Operations

In: Space Exploration
Editor: Thomas R. Reed

ISBN: 978-1-53615-032-2
© 2019 Nova Science Publishers, Inc.

Chapter 2

DEVELOPMENT OF NASA'S SMALL FISSION POWER SYSTEM FOR SCIENCE AND HUMAN EXPLORATION[*]

Marc A. Gibson[1], Lee S. Mason[1], Cheryl L. Bowman[1], David I. Poston[2], Patrick R. McClure[2], John Creasy[3] and Chris Robinson[3]
[1]National Acronautics and Space Administration
Glenn Research Center Cleveland, Ohio, US
[2]Los Alamos National Laboratory, Los Alamos, New Mexico, US
[3]National Security Complex, Oak Ridge, Tennessee, US

ABSTRACT

Exploration of our solar system has brought great knowledge to our Nation's scientific and engineering community over the past several

---

[*] This is an edited, reformatted and augmented version of NASA/TM Publication No. 2015-218460, Prepared for the 50th Joint Propulsion Conference, dated 2015.

decades. As we expand our visions to explore new, more challenging destinations, we must also expand our technology base to support these new missions. NASA's Space Technology Mission Directorate is tasked with developing these technologies for future mission infusion and continues to seek answers to many existing technology gaps. One such technology gap is related to compact power systems (>1 kWe) that provide abundant power for several years where solar energy is unavailable or inadequate. Below 1 kWe, Radioisotope Power Systems have been the workhorse for NASA and will continue, assuming its availability, to be used for lower power applications similar to the successful missions of Voyager, Ulysses, New Horizons, Cassini, and Curiosity. Above 1 kWe, fission power systems (FPSs) become an attractive technology offering a scalable modular design of the reactor, shield, power conversion, and heat transport subsystems. Near-term emphasis has been placed in the 1 to 10 kWe range that lies outside realistic radioisotope power levels and fills a promising technology gap capable of enabling both science and human exploration missions. History has shown that development of space reactors is technically, politically, and financially challenging and requires a new approach to their design and development. A small team of NASA and Department of Energy experts are providing a solution to these enabling FPS technologies starting with the lowest power and most cost-effective reactor series named "Kilopower" that is scalable from approximately 1 to 10 kWe.

INTRODUCTION

Over the past five decades numerous space reactors have been designed in the United States with little success in hardware development and no success in flight. As NASA's budget continues to be stretched over a growing mission portfolio, efficient technology development is crucial to building a healthy U.S. space nuclear program capable of producing flight systems. The Kilopower team is challenging the overly expensive technology development process used in past failures by matching a simple reactor to a smart test program with flight hardware as the end success criteria. The development strategy involves a strong test program utilizing existing facilities that will help reduce high-risk components and design assumptions through extensive testing. The program has been carefully planned to match the fidelity of the design to affordable testing capabilities and provide successive steps in the development cycle. The reactor design will advance throughout the test program with a near-term goal of a full-scale nuclear ground test, nicknamed "KRUSTY" (Kilopower Reactor Using Stirling Technology), using the flight prototypic highly enriched uranium molybdenum (UMo) core, heat pipe thermal transport systems, and Stirling power conversion. If successful, the technology demonstration will take approximately 3 years and $10 million to complete. The fidelity of the design at the end of the 3 years is anticipated to address a major portion of the technical risk associated with a flight unit with minimal invested cost. Figure 1 illustrates the flight concept of a 4-kWt reactor core coupled to a 1-kWe Stirling power conversion system and a 40-kWt core coupled to a 10-kWe conversion assembly. The 1-kWe Kilopower reactor is the baseline design for the nuclear test and will affordably address Kilopower designs ranging from 1 to 10 kWe.

The basic design approach is derived from a 2010 Small Fission Power System Feasibility Study performed by a joint NASA/Department of Energy (DOE) team in response to a request from the National Research Council (NRC) Planetary Science Decadal Survey [Ref. 1]. After reviewing options, the team selected a reference concept and evaluated the feasibility for a 10-year flight system development to support future space science missions that exceed radioisotope power system capabilities or require too much ^{238}Pu fuel. The reference concept defined in the 2010 study is similar to the current Kilopower concept, sharing the cast UMo fuel form, BeO reflector, and Na heat pipes. The baseline reactor thermal power level for the 2010 concept was 13 kWt, allowing a thermoelectric power system option at 1 kWe [Ref. 2] or a Stirling power system option at 3 kWe [Ref. 3].

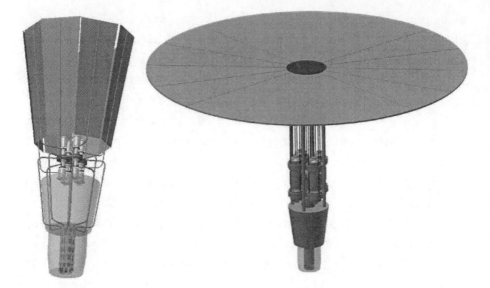

Figure 1. The 1-kWe (left) and 10-kWe (right) Kilopower systems.

SYSTEM DESIGN

Reactor

The Los Alamos National Laboratory is leading the design of the reactor, reflector, and shield that will be demonstrated in the full-scale nuclear test. The Y–12 National Security Complex is leading the material development and manufacturing of the uranium alloy core and will verify fuel selection through early material testing and manufacturing trials leading up to the full-scale fabrication. As part of the development phase, there will be numerous tests to validate the design and material selections with expected iterations leading up the final design. A solid cast UMo alloy core has been selected as the fuel because of an existing material database, simple reactor construction and operation, and most importantly, the existing infrastructure and production capabilities at Y–12 National Security Complex. The solid core works well with lower power reactors because of negligible fuel burnup and volume swelling issues that can challenge higher power reactors, which would typically incorporate cladded pin-type fuel [Ref. 4]. The solid core design provides sufficient thermal power while reducing the fuel, radial reflector, and shadow shield geometry and mass, giving the total system a higher specific power (W/kg) than other fuel forms. The baseline material for the core has been chosen to be 93 percent highly enriched U235 alloyed with 7 percent Mo by weight, and is expected to produce an optimum balance between neutronic, thermal, and metallurgical properties [Ref. 5]. Figure 2 shows the current Los Alamos National Laboratory design of a 4-kWt flight core, which provides 1 kWe with Stirling power conversion.

Beryllium oxide has been selected as the radial reflector material for its high neutron reflectance throughout the required temperature and energy spectrum as well as its mechanical and thermal properties. The reflector design is monolithic and does not incorporate control drums typical of higher power reactors, highlighting another significant and simplified design feature. For criticality safety throughout Assembly Test and Launch Operations (ATLO), only a single control rod of B₄C is needed in the center of the core to keep k_{eff} at a safe level during all expected operations and hypothetical accidents. When the reactor reaches its startup location in space or on the surface of another planet, a control mechanism will slowly remove the poison control rod and allow the reactor to start up. This benefit of controlled startup at the mission destination allows beginning of mission (BOM) power levels and overall power system life expectancy to be directly coupled to mission timeline requirements. The flight system could utilize a number of control rod options that could be specifically tailored to mission requirements. For instance, the simplest control rod design, as baselined, would perform only one movement at the beginning of the mission allowing the reactor to start up and load follow the power conversion system. This method allows the natural degradation of the core temperature as fuel is spent and neutronic behavior changes over time. Conservative thermal degradation of the 4-kWt core using this method are estimated to be 3 K/year with 0.1 percent fuel burnup over 15 years. An alternative approach is to use an active control rod that adds reactivity as needed throughout the mission to keep the reactor temperature and power constant. Using active control, the reactor can provide constant power for several hundred years at the 4 kWt level due to very little fuel burnup and needed reactivity insertion. These simple control rod and startup methods are an important design feature needed for missions that may require the reactor to start up under minimum power at a location where solar energy is limited and battery power is necessary. Initial estimates of the required power for startup is 1 amp hr at 28 V of direct current.

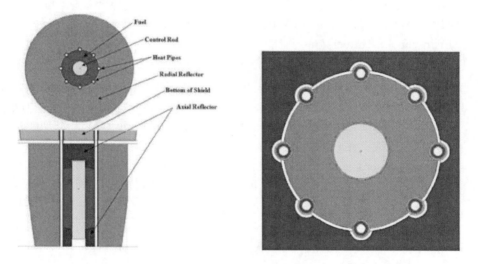

Figure 2. Flight design of the Kilopower 4-kWt core and reflector assembly.

HEAT TRANSPORT AND POWER CONVERSION

Thermal energy from the core is transported to the power convertors via sodium heat pipes. Heat generated from the fission reactions is conducted through the core and into the heat pipe evaporator and vaporizes the sodium liquid. The sodium vapor travels up the heat pipe where it can be accepted by the Stirling convertors at the condenser interface. As the sodium vapor releases its latent heat and condenses back to the liquid phase, the wick pumps it back to the evaporator where the cycle continues. Alloy 230 is the baseline envelope material for the heat pipes because of its known compatibility and prior experience with sodium as well as its high temperature strength and creep resistance. This passive thermal transport operates solely on thermal energy and requires no electrical power for pumping. This is an important design feature, which reduces the parasitic losses of the power system and simplifies system startup and control.

High-efficiency free-piston Stirling convertors have been baselined for the initial designs to increase system performance and provide high specific power. Their use benefits from existing flight development of the Sunpower, Inc., 80-We Advanced Stirling Convertor (ASC) as well as recent successful technology demonstrations of both 1- and 6-kWe convertors developed by Sunpower, Inc., for NASA under the current Nuclear Systems Program. The Stirling engine heat acceptor is conductively coupled to the sodium heat pipe condenser and uses the thermal energy from the reactor to thermodynamically drive the power piston and linear alternator. The Stirling convertors in both the 1- and 10-kWe Kilopower designs are arranged in the vertical dual-opposed configuration allowing easy power scaling while minimizing the shield half angle and mass. Thermoelectric conversion has been studied as an alternate power conversion technology that offers simplicity and additional redundancy but requires significantly more thermal power from the reactor due to its lower efficiency [Ref. 2]. NASA studies are currently looking at these two power conversion options for future development and mission use. The preliminary baseline design (Figure 1) uses eight 125-We ASC-style convertors in a dual-opposed configuration with mating hot ends. Coupling the hot ends of all eight convertors is a conduction plate that allows redundant heat paths in the event of a heat pipe or convertor failure. This hot-end assembly increases reliability but also reduces mass using a unified insulation package and fewer components. Future Stirling designs will likely incorporate a shared expansion space between engine pairs similar to the current 6-kWe Sunpower, Inc., design [Ref. 6].

The Stirling engines must reject their waste heat to space at the optimum temperature in order to establish a balance between the conversion efficiency and radiator mass. This optimum temperature is not equal for the 1- and 10-kWe systems but does fit well within the operating range for water-based heat pipes. Titanium-water heat pipes are baselined for the

Kilopower systems and have already been through a significant development cycle with numerous successful designs and tests. The titanium-water heat pipes will transfer the waste heat from the Stirling engines to a radiator fin where it can then be rejected to space.

MASS AND PERFORMANCE SCALING
OF KILOPOWER SYSTEMS

Preliminary mass estimates were generated for Kilopower systems from 1 to 10 kWe, assuming the same basic reactor design parameters defined in the 2010 NRC feasibility study. The results are presented in Table I. The 1-kWe system allows the heat pipes to be clamped to the outside perimeter of the core, while the larger systems will require the heat pipes to be inserted into the core. Increasing power output requires a slightly larger core and more heat pipes, but no major changes to the basic design approach. The radiator area is based on one-sided heat rejection to a 200 K thermal sink with 10 percent area margin. The systems could employ fixed radiators for planetary spacecraft applications or deployable radiators for Mars surface applications. The stowed dimensions are provided for both options. The mass breakouts are shown for the reactor, shield, and balance of plant. As indicated, the system specific power improves with increasing power level from about 2.5 W/kg at 1 kWe to 6.5 W/kg at 10 kWe.

The system concepts are meant to support both science and human exploration uses [Ref. 7]. For science, a number of mission concepts were studied during the NRC Planetary Science Decadal Survey with projected power requirements greater than 500 W [Ref. 8]. These missions may include orbiters, landers, multiple science targets, expanded instrument suites, in situ data analysis, high-rate communications, and/or electric propulsion. Human outpost missions to the Moon and Mars have generally required tens of kilowatts to accommodate landers, habitats, In-Situ Resource Utilization (ISRU) plants, rover recharging, communication relays, and science packages. Until recently, the preferred approach was to deliver a large, centralized 40-kWe fission power plant that could meet the mission requirements with margin for growth. The drawback was the large landed mass, complex installation, and lack of system-level redundancy. The 1- to 10-kWe fission power option could be used for human precursor missions to demonstrate ISRU production or on human outpost missions where compact packaging, modularity, and redundancy are favored over growth capacity. An additional application for the small fission system that is being considered for Mars surface missions is to provide a portable utility pallet for remote charging of long-distance piloted rovers.

Table I. Kilopower System Options

User Power, kWe	1	3	5	7	10
Power System Characteristics					
Reactor thermal power, kWt	4.3	13.0	21.7	30.3	43.3
Radiator area, m^2	3.2	9.6	13.5	17.1	20.0
Stowed diameter, m	1.1	1.2	1.3	1.4	1.5
Stowed height (with fixed radiator), m	3.0	4.9	5.9	6.7	7.3
Stowed height (with deployable radiator), m	N/A	2.2	2.7	3.0	3.3
Mass Summary, kg					
Reactor (UMo core, BeO reflector, Na heat pipes)	136	175	198	215	235
Shield (LiH/W, 25 krad and 10^{11} n/cm^2 at 10 m separation)	148	272	364	443	547
Balance-of-plant (Stirling engines, radiator, electric controls, structure)	122	304	449	589	763
Power system, kg	406	751	1011	1246	1544
Specific power, W/kg	2.5	4.0	4.9	5.6	6.5

THE DEVELOPMENT PROGRAM

Early demonstration of the Kilopower technology was proven with the DUFF (Demonstration Using Flattop Fission) experiment in 2011 [Refs. 9 and 10]. This test not only showed that nuclear testing of space reactors and their subcomponents can be affordably tested but that there are existing nuclear facilities that can be utilized for development programs [Ref. 11]. This proof-of-concept test successfully showed that reactors could be coupled to Stirling power convertors via heat pipes and produce electricity. Building on the DUFF experiment, the Kilopower team has planned a follow-on 3-year development program that is designed to reduce the technical risk associated with the flight unit by addressing many first- and second-order design uncertainties. The 3-year program is expected to start at the beginning of NASA fiscal year 2015 and end with the KRUSTY test by the start of fiscal year 2018. In order to complete the program, many tests will be performed to determine if the preliminary design will function as expected with the understanding that modifications will be necessary to evolve towards the final design. These tests will include material testing, simple thermal mockup tests, higher fidelity component and system tests, and end with the full-scale nuclear test. Using this strategy, early first-order design flaws will be exposed with little investment and plenty of time for design modifications. All mechanical, thermal, and power conversion systems will be completely verified before nuclear testing is commenced.

MATERIAL TESTING AND FABRICATION

Neutronic worth is a key driver for selecting a reactor fuel. Chemical and structural stability are, however, very important design factors as well. It cannot be overemphasized that the fuel selection trade space is greatly expanded by designing the fission system to a very low fuel burnup. The gas generation and fuel swelling that result from advanced radiological stages are crucial drivers in most fuel systems but are essentially nonfactors in the design trade-space discussed here. Fabricability, structural stability, chemical stability as well as neutronic worth were the discriminators used for fuel selection in this design concept. Uranium oxides or uranium nitrides were not selected since the additional volume stability is not needed in low burnup applications. The selection of an UMo alloy over uranium with low alloying additions was made to take advantage of the higher temperature strength, wider range of microstructural stability, and the relatively larger physical property data associated with UMo alloys. The Y–12 National Security Complex, NASA Glenn, and Los Alamos National Laboratory have conducted extensive reviews of data collected for the past five decades regarding UMo fuel properties. The 2009 UMo Fuels Handbook [Ref. 12] provides the thermophysical and much of the mechanical property data required for conceptual design of the compact fission power system (FPS). Absent from the fuels handbook is time-dependent mechanical property data, or creep data, which will be needed for this design but is not relevant in typical fuel pellet systems. Scrutiny of older literature suggests that, although the creep strength of UMo alloys is very low at these temperatures, it will be sufficient for this design [Ref. 13]. These material selection assumptions will be verified by casting and testing a lot of prototypic depleted UMo rods. Samples will be tested for compressive creep strength, tensile creep strength, impact excitation modulus, and diffusion rates relative to pure molybdenum after characterizing the chemistry and grain size of the rod.

THERMAL PROTOTYPES

Each of the Kilopower components will start with thermal testing to verify analytical models and provide test and assembly knowledge that will be useful leading up to the nuclear demonstration. Starting with the reactor core, the heat transfer mechanisms associated with the thermal coupling and temperature drop from the core to the sodium heat pipe evaporator will be demonstrated using a full-scale stainless steel 316L core, a custom graphite heater, eight alloy 230/sodium heat pipes, 316L core containment can, mock axial reflectors, and several heat pipe clamping designs. This early prototype will address several first-order design features that can easily be modified to improve performance and eventually be downselected to carry forward through the test program. The assembly has been specially designed to not only test heat transfer and mechanical coupling but will also verify assembly operations for KRUSTY integration. Figure 3 shows the thermal prototype assembly and the individual components intended for early testing. The

316L stainless steel was chosen as the surrogate material for many of the components as a cost-effective way to develop early material creep models that could verify modeling tools and creep behavior of the design. The surrogate material physical properties are very well documented and are similar enough to many of the published UMo alloy properties that results should be useful on numerous first-order effects. At the conclusion of the material and thermal prototype tests, a preliminary design review will be held to determine if the design is ready to move forward to casting a depleted UMo core for further nonnuclear testing. The thermal prototype testing will verify the following:

(1) High-temperature creep modeling using known 316L material properties
(2) Heater design and thermal behavior at nominal Kilopower operating temperatures and power levels
(3) Core to heat pipe contact resistance and thermal performance
(4) MLI (Multi-Layer Insulation) assembly and thermal performance
(5) Sodium heat pipe performance using gas calorimetry
(6) Heat pipe to core clamping mechanisms
(7) Core containment can geometry and mechanical design
(8) Assembly processes for KRUSTY and ATLO

STIRLING POWER CONVERSION AND SODIUM HEAT PIPE TESTING

The ASC power conversion units from the Advanced Stirling Radioisotope Generator (ASRG) program will be used for initial power conversion testing starting in fiscal year 2015. Two convertors will first be tested separately to reverify individual performance followed by coupling them as a pair to the Kilopower hot-end conduction plate and sodium heat pipes. Internal modifications to the convertors are expected to boost performance by 20 percent for Kilopower testing and take advantage of the higher power fission heat source. With eight sodium heat pipes and only two convertors, the remaining convertor assemblies will be replaced with thermal simulators to mimic the response of the Stirling convertors. The conduction plate will be the coupling mechanism between the eight heat pipes, six thermal simulators, and two Stirling convertors. The assembly will initially be tested in vacuum with the stainless steel core to verify system performance and thermal integrity.

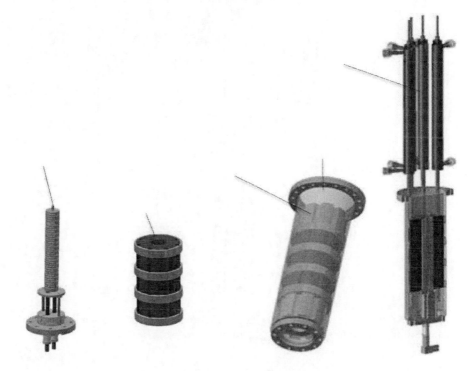

Figure 3. Thermal prototype assembly and test components.

Figure 4 illustrates the sodium heat pipe array, hot-end conduction plate, and Stirling convertors/ simulators. This test will address the following:

(1) System-level integration and assembly processes
(2) Steady-state and transient system performance
(3) Total conversion efficiency including thermal losses
(4) Flight prototypic sodium heat pipe performance

(5) System response and mechanical integrity to induced heat pipe failure System response to simulated convertor failure

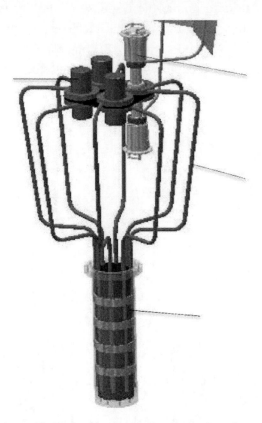

Figure 4. Kilopower Assembly Test with core, sodium heat pipes, hot-end conduction plate, Stirling convertors, and Stirling thermal simulators.

DEPLETED UMO CORE

Once the preliminary design review is complete, the engineering drawings will be released to produce a depleted UMo core at the Y–12 National Security Complex. Y–12 will build the core to print and ship to the NASA Glenn Research Center where further testing will take place. At the time of delivery, the Kilopower system will have been fully checked out and ready for integration with the depleted uranium core. This test assembly is identical to the Stirling power conversion and sodium heat pipe configuration in Figure 4 and will further the design fidelity by verifying:

(1) The UMo interface requirements between the heat pipes, insulation, and clamping mechanisms
(2) Mechanical design integrity during multiple thermal cycles and potential core deformation
(3) Thermal expansion of the core and its effects on the system
(4) Material creep of the UMo core under induced design stresses
(5) Diffusion of the UMo with interface materials
(6) Thermal vacuum acceptance criteria

HIGHLY ENRICHED UMO CORE AND THE KRUSTY TEST

At completion of the NASA thermal vacuum testing of the depleted uranium core, a final design review will be held to determine if the highly enriched uranium core is ready to begin fabrication. This final design review will accumulate all design revisions throughout the testing program and release the final design drawings to Y–12 National Security Complex for fabrication. When complete, the highly enriched uranium core will be shipped to the Device Assembly Facility (DAF) where it will integrate into the test assembly at the proper time. The full-scale nuclear testing will be performed with the Comet criticality machine at the DAF. The flight prototypic assembly for the KRUSTY test will be identical to the depleted uranium core test leaving only the nuclear design to be verified. Figure 5 illustrates the integrated test assembly with Comet depicting the two extreme reflector positions using the lift table. Comet will make the test assembly go critical by raising the radial reflector around the core and provide the necessary reactivity to create the 4-kWt steady-state thermal power. The test assembly will incorporate a custom vacuum chamber specifically designed to fit on top of Comet and provide the relevant space environment throughout the nuclear testing. The KRUSTY test will verify and/or demonstrate:

(1) Reactor startup operations
(2) Excess reactivity needed to meet Kilopower thermal power and temperature requirements
(3) Integral nuclear cross sections and temperature dependence
(4) Reactor load following to Stirling convertor demands
(5) ATLO assembly procedures
(6) Steady state and transient differences between electrical and nuclear heat sources
(7) Temperature feedback mechanisms and dynamic response
(8) Operational stability for follow-on engineering unit nuclear tests
(9) Nuclear design tools such as FRINK and MRPLOW

CONCLUSION

Developing a small fission power system for NASA's science and human exploration is an endeavor worth taking with potential to open up a new class of missions not currently achievable with radioisotope and solar power sources. An affordable approach to addressing many engineering risks of a future flight development program have been proposed to take approximately 3 years and $10 million using a test plan that progresses through increasing levels of hardware fidelity leading up to a full nuclear ground test nicknamed "KRUSTY" (Kilopower Reactor Using Stirling TechnologY). This development project of Kilopower will provide extensive science and engineering data not attained in the last five decades of U.S. space reactor programs.

Starting with the lower power 4-kWt reactor core for the first nuclear demonstration is extremely important to keeping development costs at an affordable level. Nuclear testing costs are directly proportional to reactor thermal power and the 4-kWt design allows the testing to take place at existing facilities under current regulations and licensing at the Nevada test site. By design, the lower power demonstration offers a subscale test of a 10 kWe capability, adding considerable value to both science and human exploration needs and paving the way for future higher power systems.

Successful nuclear testing of the Kilopower reactor will help fill the existing technology gap of compact power systems in the 1 to 10 kWe range enabling new higher power NASA science and human exploration missions.

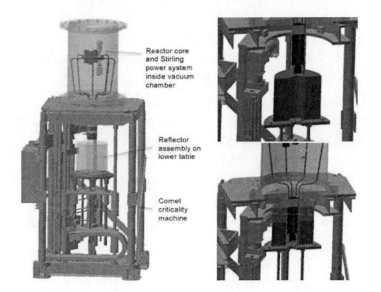

Figure 5. KRUSTY test assembly integrated with the Comet criticality machine.

REFERENCES

[1]   Committee on the Planetary Science Decadal Survey: *Vision and Voyages for Planetary Science in the Decade 2013–2022*. National Research Council, The National Academies Press, Washington DC, 2011.

[2]   Mason, L., et al.: *A Small Fission Power System for NASA Planetary Science Missions*. NASA/TM—2011-217099, 2011.

[3]   Mason, L.; and Carmichael, C.: *A Small Fission Power System With Stirling Power Conversion for NASA Science Missions*. NASA/TM—2011-217204, 2011.

[4]   Poston, D.I., et al.: *Notional Design of the Kilopower Space Reactor*. NETS–2014–2007, 2014.

[5]   Creasy, J.; and Bowman, C.: *Fuel Selection and Development for Small Fission Power Systems.* NETS–2014–2006, 2014.

[6]   Geng, S.M., et al.: Development of a 12 kWe Stirling *Power Conversion Unit for Fission Power Systems-Status Update.* NETS–2014–2004, 2014.

[7]   Mason, L.; Gibson, M.A.; and Poston, D.I.: *Kilowatt-Class Fission Power Systems for Science and Human Precursor Missions.* NASA/TM—2013-216541, 2013.

[8]   *Planetary Science Decadal Survey Mission & Technology Studies.* The National Academies Press, Washington DC, 2015. http://sites. nationalacademies.org/SSB/SSB_059331 Accessed Feb. 17, 2015.

[9]   Poston, D.I., et al.: *The DUFF Experiment-What was Learned?* NETS–2013–6967, 2013.

[10]   Gibson, M.A., et al.: *Heat Pipe Powered Stirling Conversion for the Demonstration Using Flattop Fission (DUFF) Test.* NASA/TM—2013-216542, 2013.

[11]   McClure, P.R.; and Holt, J.L.: *The Use of the Nevada National Security Site as a Reactor Test Center.* NETS–2014, NASA Stennis Space Center, MS, 2014.

[12]   Rest, J., et al.: *UMo Fuels Handbook.* ANL-09/31, 2009.

[13]   Nichols, R.W.: *Uranium and its Alloys. Nuclear Engineering*, 1957, pp. 355–365.

In: Space Exploration
Editor: Thomas R. Reed

ISBN: 978-1-53615-032-2
© 2019 Nova Science Publishers, Inc.

Chapter 3

SPACE EXPLORATION: DOE COULD IMPROVE PLANNING AND COMMUNICATION RELATED TO PLUTONIUM-238 AND RADIOISOTOPE POWER SYSTEMS PRODUCTION CHALLENGES[*]

United States Government Accountability Office

ABBREVIATIONS

ASRG	Advanced Stirling Radioisotope Generator
ATR	Advanced Test Reactor
DOE	Department of Energy
eMMRTG	enhanced Multi-Mission Radioisotope Thermoelectric Generator

---

[*] This is an edited, reformatted and augmented Accessible version of the United States Government Accountability Office Report to Congressional Requesters, Publication No. GAO-17-673, dated September 2017.

GPHS	general purpose heat source
HEO	Human Exploration and Operations Mission Directorate
HFIR	High Flux Isotope Reactor
INL	Idaho National Laboratory
kg	Kilogram
LANL	Los Alamos National Laboratory
MMRTG	Multi-Mission Radioisotope Thermoelectric Generator
NASA	National Aeronautics and Space Administration
ORNL	Oak Ridge National Laboratory
PSD	Planetary Science Division
Pu-238	plutonium-238
RPS	radioisotope power system
Supply Project	Pu-238 Supply Project

WHY GAO DID THIS STUDY

NASA uses RPS to generate electrical power in missions in which solar panels or batteries would be ineffective. RPS convert heat generated by the radioactive decay of Pu-238 into electricity. DOE maintains a capability to produce RPS for NASA missions, as well as a limited and aging supply of Pu-238 that will be depleted in the 2020s, according to NASA and DOE officials and documentation. With NASA funding, DOE initiated the Pu-238 Supply Project in 2011, with a goal of producing 1.5 kg of new Pu-238 per year by 2026. Without new Pu-238, future NASA missions requiring RPS are at risk.

GAO was asked to review planned RPS and Pu-238 production to support future NASA missions. This chapter (1) describes how NASA selects RPS for missions and what factors affect RPS and Pu-238 demand; and (2) evaluates DOE's progress and challenges in meeting NASA's RPS and Pu-238 demand. GAO reviewed NASA mission planning and DOE program documents, visited two DOE national laboratories involved in making new Pu-238 or RPS work, and interviewed agency officials.

WHAT GAO RECOMMENDS

GAO is making three recommendations, including that DOE develop a plan with milestones and interim steps for its Pu-238 and RPS production approach, and that DOE assess the long-term effects of known production challenges and communicate these effects to NASA. DOE concurred with GAO's recommendations.

WHAT GAO FOUND

The National Aeronautics and Space Administration (NASA) selects radioisotope power systems (RPS) for missions primarily based on the agency's scientific objectives and mission destinations. Prior to the establishment of the Department of Energy's (DOE) Supply Project in fiscal year 2011 to produce new plutonium-238 (Pu-238), NASA officials said that Pu-238 supply was a limiting factor in selecting RPS-powered missions. After the initiation of the Supply Project, however, NASA officials GAO interviewed said that missions are selected independently of decisions on how to power them. Once a mission is selected, NASA considers power sources early in its mission review process. Multiple factors could affect NASA's demand for RPS and Pu-238. For example, high costs associated with RPS and missions can affect the demand for RPS because, according to officials, NASA's budget can only support one RPS mission about every 4 years. Expected technological advances in RPS efficiency could reduce NASA's demand for RPS and Pu-238.

DOE has made progress in reestablishing Pu-238 production to meet NASA's future demand to fuel RPS and has identified challenges to meeting its production goals. Specifically, since the start of the Supply

Project, DOE has produced 100 grams of Pu-238 and expects to finalize production processes and produce interim quantities by 2019. However, DOE has also identified several challenges to meeting the Supply Project goal of producing 1.5 kilograms (kg) of new Pu-238 per year by 2026. DOE officials GAO interviewed said that DOE has not perfected the chemical processing required to extract new Pu-238 from irradiated targets to meet production goals. These officials also said that achieving the Pu-238 production goal is contingent on the use of two reactors, but only one reactor is currently qualified for Pu-238 production while the second reactor awaits scheduled maintenance. Moreover, while DOE has adopted a new approach for managing the Supply Project and RPS production—based on a constant production approach—the agency has not developed an implementation plan that identifies milestones and interim steps that can be used to demonstrate progress in meeting production goals and addressing previously identified challenges. GAO's prior work shows that plans that include milestones and interim steps help an agency to set priorities, use resources efficiently, and monitor progress in achieving agency goals. By developing a plan with milestones and interim steps for DOE's approach to managing Pu-238 and RPS production, DOE can show progress in implementing its approach and make adjustments when necessary. Lastly, DOE's new approach to managing the Supply Project does not improve its ability to assess the potential long-term effects of challenges DOE identified, such as chemical processing and reactor availability, or to communicate these effects to NASA. For example, DOE officials did not explain how the new approach would help assess the long-term effects of challenges, such as those related to chemical processing. Under *Standards for Internal Control in the Federal Government*, agencies should use quality information to achieve objectives and to communicate externally, so that external parties can help achieve agency objectives. Without the ability to assess the long-term effects of known challenges and communicate those effects to NASA, DOE may be jeopardizing NASA's ability to use RPS as a power source for future missions.

September 8, 2017

The Honorable Lamar Smith
Chairman
Committee on Science, Space,
and Technology House of Representatives

The Honorable Brian Babin
Chairman
Subcommittee on Space
Committee on Science, Space, and Technology
House of Representatives

The National Aeronautics and Space Administration (NASA) has long used radioisotope power systems (RPS) to generate reliable electrical power and heat energy for long-duration space missions. RPS produce power by converting heat from the natural radioactive decay of plutonium-238 (Pu-238) into electricity and can operate where solar panels or batteries would be ineffective or impossible to use, such as in deep space or in shadowed craters.[1] RPS also have the advantage of being able to operate continuously and provide power for more than a decade. Currently, a single RPS unit is being used to power the Mars Science Laboratory, also known as Curiosity, NASA's unmanned robotic surface rover that has been exploring the planet Mars since 2012.

The Department of Energy (DOE) and its predecessor agencies have been providing Pu-238 and fabricating RPS for NASA and other federal agencies for more than 5 decades. Historically, Pu-238 was produced domestically or was purchased from Russia. Domestic Pu-238 production ended in 1988, and DOE has not purchased material from Russia since 2009.[2] As a result, supplies of available Pu-238 to support new missions

---

[1] For the purposes of this report, unless otherwise noted, Pu-238 is defined as Pu-238 oxide, also known as "heat-source" plutonium oxide or "bulk-oxide", and is the form used to power RPS. Pu-238 isotope is a precursor to Pu-238 oxide.

[2] Pu-238 was last produced at DOE's Savannah River Site in South Carolina in 1988, using nuclear reactors that have since been shut down. Pu-238 production at the site was facilitated from

have diminished. Because of a limited availability of Pu-238, the National Academy of Sciences expressed concern about future missions in the *Vision and Voyages for Planetary Science in the Decade 2013-2022* survey report, which identifies the science community's highest priority space exploration interests,[3] because many of the highest priority missions identified in the report can only be enabled by RPS.

According to DOE documents and agency officials, DOE currently maintains about 35 kilograms (kg) of Pu-238 isotope designated for NASA missions,[4] about half of which meets power specifications for spaceflight.[5] This supply, however, could be exhausted within the next decade based on NASA's solar system exploration plans. Specifically, NASA plans to use about 3.5 kg of Pu-238 isotope for one RPS for the Mars 2020 mission. NASA could use an additional 10.5 kg of Pu-238 isotope for the New Frontiers #4 mission if it were to use three RPS, a decision that NASA officials said the agency expects to make in July 2019 to support a 2025 launch window. If DOE's existing Pu-238 supply is used for these two missions, according to DOE documentation, NASA will be forced to eliminate RPS as a power source for future missions, delaying future missions that require RPS until DOE produces or acquires more Pu-238. New Pu-238 can be blended with existing Pu-238 that does not meet power specifications so that the blended Pu-238 can be used for future NASA missions.

Under the authority of the Atomic Energy Act of 1954, DOE maintains the nation's capability to support the development, production, and safety of RPS used in NASA's space exploration missions.[6] Three DOE national laboratories—Idaho National Laboratory (INL), Oak Ridge National

---

byproducts of nuclear weapons production. Chemical processing activities at the site were scheduled for shut-down following the completion of their mission to prepare Cold War legacy nuclear materials.

[3] National Research Council of the National Academies, *Vision and Voyages for Planetary Science in the Decade 2013-2022* (Washington, D.C.: National Academy of Sciences, 2011).

[4] DOE manages a separate allocation of Pu-238 for national security purposes apart from the 35 kg set aside for NASA.

[5] Pu-238 must meet a minimum specification of 1,952 watts of heat, among other characteristics, to be used for spaceflight.

[6] Pub. L. No. 83-703, 68 Stat. 919 (codified as amended at 42. U.S.C. §§ 2011-2297-h13).

Laboratory (ORNL), and Los Alamos National Laboratory (LANL)—are currently involved in RPS production. In 2011, with funding provided by NASA, DOE initiated the Pu-238 Supply Project (Supply Project) in order to reestablish the capability to domestically produce Pu-238. Since 2011, DOE has produced approximately 100 grams of Pu-238 isotope under the Supply Project.[7] DOE identified an interim goal of producing 300 to 500 grams of new Pu-238 per year by 2019, and in 2010, it established a goal of producing 1.5 kg of new Pu-238 per year—considered full production— by 2023, at the earliest.

In 2009, the National Academy of Sciences[8] reported that NASA has been making mission-limiting decisions for some time because of a limited supply of Pu-238 and that NASA has been eliminating RPS as an option for some missions and delaying other missions that require RPS until DOE can reestablish production of Pu-238. In the National Aeronautics and Space Administration Transition Authorization Act of 2017,[9] Congress required NASA and the Office of Science and Technology Policy to conduct an analysis of, among other things, the risks associated with NASA's ability to carry out planned, high-priority robotic missions in the solar system and other surface exploration activities beyond low-Earth orbit, resulting from a lack of adequate radioisotope power system material or Pu-238.

You asked us to examine NASA's process for considering and selecting power sources for missions, in particular the use of RPS, and to review DOE's ability to maintain the necessary infrastructure and workforce for RPS and Pu-238 production. This chapter (1) describes how NASA selects RPS for missions and what factors affect RPS and Pu-238 demand; and (2) evaluates DOE's progress in meeting NASA's RPS and Pu-238 demand and what challenges, if any, DOE faces in meeting the demand.

---

[7] Initial samples of this newly produced Pu-238 did not meet space flight specifications because they contained impurities. These samples can, however, be blended and used with existing Pu-238, according to DOE officials.

[8] National Research Council of the National Academies, *Radioisotope Power Systems: An Imperative for Maintaining U.S. Leadership in Space Exploration* (Washington, D.C.: National Academy of Sciences, 2009).

[9] Pub. L. No. 115-10, § 515(b)(2), 131 Stat. 18, 53.

To describe how NASA selects RPS for missions and what factors affect RPS and Pu-238 demand, we reviewed documentation related to how NASA's mission procedural requirements were considered during planning for recent missions that considered or used RPS as a power source.[10] We also interviewed officials from the Planetary Science Division (PSD) of NASA's Science Mission Directorate and from the Human Exploration and Operations Mission Directorate (HEO).

To evaluate DOE's progress in meeting NASA's RPS and Pu-238 demand, and what challenges, if any, DOE faces in meeting the demand, we reviewed documentation related to DOE's efforts to develop the Supply Project and to DOE's RPS production process. We also interviewed officials from DOE's Office of Nuclear Energy and DOE's national laboratories involved in RPS work—INL, LANL, and ORNL—and conducted site visits to ORNL, the laboratory responsible for the Supply Project, and INL, the laboratory primarily responsible for overseeing RPS production. Finally, we compared DOE's efforts to collect and assess quality information about challenges associated with RPS and Pu-238 production and to communicate these challenges against criteria outlined in federal internal control standards.[11] We also evaluated DOE's management approach for RPS and Pu-238 production against key management practices established in prior GAO work.[12] For more detailed information on our methodology, see appendix I.

We conducted this performance audit from March 2016 to September 2017 in accordance with generally accepted government auditing standards. Those standards require that we plan and perform the audit to obtain sufficient, appropriate evidence to provide a reasonable basis for our findings and conclusions based on our audit objectives. We believe that the

---

[10] National Aeronautics and Space Administration, *NASA Space Flight Program and Project Management Requirements*, NPR 7120.5D (Sept. 28, 2011).

[11] GAO, *Standards for Internal Control in the Federal Government*, GAO-14-704G (Washington, D.C.: Sept. 10, 2014).

[12] GAO, *Defense Health Care Reform: Actions Needed to Help Ensure Defense Health Agency Maintains Implementation Progress*, GAO-15-759 (Washington, D.C.: Sept. 10, 2015), and *Biobased Products: Improved USDA Management Would Help Agencies Comply with Farm Bill Purchasing Requirements*, GAO-04-437 (Washington, D.C.: Apr. 7, 2004).

evidence obtained provides a reasonable basis for our findings and conclusions based on our audit objectives.

BACKGROUND

This section provides information on the use of RPS in NASA space missions, NASA's PSD mission portfolio and mission classes, DOE's role in RPS production, DOE's Pu-238 Supply Project, RPS production across DOE's national laboratories, and NASA's funding for RPS production.

Use of RPS in NASA Space Missions

RPS are long-lived sources of spacecraft electrical power and heating that are rugged, compact, highly reliable, and relatively insensitive to radiation and other effects of the space environment, according to NASA documentation (see figure 1). Such systems can provide spacecraft power for more than a decade and can do so billions of miles from the sun. In addition to providing electricity, heat generated by Pu-238 contained in the RPS is used to keep spacecraft instruments and components warm in the frigid environments of deep space. Waste heat is rejected into the environment via radiator fins that are attached to the RPS. Twenty-seven U.S. missions have used RPS over the past 5 decades. RPS produce electrical power by converting the heat generated by the natural radioactive decay of Pu-238. Pu-238 is the best possible fuel for RPS because it emits radiation that is easily shielded for the spacecraft, is producible in sufficient quantities from available material, and can be made into stable chemical forms that will not be taken up into the environment if accidentally released. The current RPS design, the Multi-Mission Radioisotope Thermoelectric Generator (MMRTG), converts heat given off by Pu-238 into about 120 watts of electrical power at the beginning of its life—a 6 percent power conversion efficiency.[13] One MMRTG contains 32 general purpose heat source (GPHS) fuel clads, which are pressed Pu-238 pellets encapsulated in iridium. Four fuel clads are encased in one GPHS module, and eight of these modules are used to fuel one RPS under the current MMRTG design.

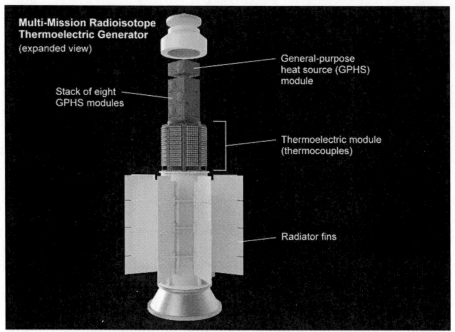

Source: National Aeronautics and Space Administration. | GAO-17-673.

Figure 1. Expanded View of the Multi-Mission Radioisotope Thermoelectric Generator (MMRTG).

NASA's PSD Mission Portfolio and Mission Classes

NASA's PSD science portfolio includes a wide array of missions that seek to address a variety of scientific objectives. PSD's scientific objectives aim to answer many questions about the solar system, from how life began to how the solar system is evolving, through the study of the planets and small bodies that inhabit the solar system. NASA's missions and associated

---

[13] Over time, as Pu-238 degrades, it gives off less heat, and therefore the RPS produces fewer watts.

mission objectives seek to address PSD's scientific objectives. Some potential mission objectives include returning samples from the surface of a comet, more intensive observations of the planets and previously unobserved small bodies, and planet surface exploration using unmanned robotic surface rovers. Scientific and mission objectives influence the types of equipment needed for the mission, including the mission's power source.

While NASA has used RPS in support of answering PSD's scientific objectives, NASA has not recently used RPS to pursue scientific objectives for other divisions or directorates, such as HEO, which oversees human spaceflight missions. The missions NASA conducts from PSD's portfolio are of two types—directed and competed. Directed missions are planned and carried out by NASA or another project management organization to accomplish a specific set of priority objectives, while competed missions are competitively awarded and carried out by principal investigators under NASA oversight.[14] NASA offers competed missions through announcements of opportunity, which specify scientific objectives and mission requirements, including potential mission destinations and projected launch dates, and what equipment will be made available, such as RPS. According to NASA officials, missions in NASA's PSD portfolio are generally classified in three ways:

- Flagship. Flagship missions are the largest and most expensive of NASA's mission classes, costing $2 billion or more, and are given the highest priority for resources, including funding, infrastructure, and launch support. Flagship missions are directed by NASA to fulfill specific needs or gaps in scientific knowledge. Past Flagship missions that have used RPS include Galileo, Cassini, and Curiosity, and Mars 2020 is a planned Flagship mission using RPS.
- New Frontiers. Added to NASA's budget in 2003, New Frontiers missions are competed missions that focus on enhancing our understanding of the solar system and that have a development cost

---

[14] Principal investigators provide scientific and technical leadership for proposed research associated with competed missions.

cap of $850 million.[15] New Frontiers missions are awarded to principal investigators. To date, there have been three New Frontiers missions—New Horizons, which is RPS-powered and is observing Pluto and the outer solar system; Juno, which uses solar power to observe Jupiter; and the Origins-Spectral Interpretation-Resource Identification Security-Regolith Explorer (OSIRIS-REx), which uses solar power and batteries and is to return samples from an asteroid back to Earth.

- Discovery. Missions in the Discovery program, which was started in 1992, are also competed, have a development cost cap of $450 million to $500 million according to NASA officials and documentation, and have the goal of enhancing our understanding of the solar system. The Discovery program utilizes many smaller missions with fewer resources and shorter development times. Discovery missions have never been powered by RPS.

DOE's Role in RPS Production

The Atomic Energy Act of 1954 authorizes DOE to provide systems that meet the special nuclear material needs of other federal agencies, and under an agreement with NASA, DOE is responsible for maintaining our nation's capability to support the development, production, and safety of NASA's space exploration missions that use RPS.[16] According to DOE officials, DOE's Office of Nuclear Energy, Nuclear Infrastructure Programs, manages and oversees RPS production and the Supply Project and coordinates with NASA to ensure DOE can meet NASA's mission requirements.[17] This includes designing, developing, fabricating, testing, and delivering RPS to

---

[15] Mission cost caps are in fixed fiscal year 2015 dollars and do not include certain costs, such as those related to the launch vehicle and operations.

[16] DOE also manages a separate allocation of existing Pu-238 for national security applications.

[17] According to DOE officials, DOE conducted a reorganization effort in September 2016 that resulted in shifting RPS production responsibilities from what was the Office of Nuclear Energy's Office of Space and Defense Power Systems to its Nuclear Infrastructure Programs office.

meet NASA's overall systems requirements, specifications, and schedules. DOE also maintains RPS production infrastructure to sustain capabilities between NASA missions and to support mission planning.

DOE's Pu-238 Supply Project

DOE restarted efforts to establish domestic Pu-238 production at the end of 2011, under its Supply Project, to provide fuel for RPS. DOE's goal is to reach a full Pu-238 production rate of 1.5 kg per year by 2023, at the earliest, with a late completion date of 2026. Before this full production rate is achieved, DOE established an interim production rate of 300 to 500 grams per year by 2019 in order to ensure an adequate supply of Pu238 for NASA's near-term missions. Until March 2017, DOE divided work associated with the Supply Project into discrete segments. The segmented management approach associated with the Supply Project was a short-term, incremental approach to managing a program with uncertain funding levels, according to DOE officials. According to DOE officials, the segmented approach was intended to allow DOE to establish and reach near-term Supply Project goals while maintaining a base level of trained staff as the project developed from its early stages to its full production rate. In March 2017, DOE officials changed how they manage the Supply Project, discontinuing the segmented approach. The new management approach for the Supply Project is to align with how DOE expects to manage RPS production, according to DOE officials.

DOE's Technical Integration Office, based at INL, coordinates with other DOE laboratories on Supply Project work. As shown in figure 2, the Supply Project involves a number of steps across several DOE national laboratories, including INL and ORNL. The production of new Pu-238 under the Supply Project begins with a shipment of neptunium-237 (neptunium) isotope from INL to ORNL.[18] The neptunium is blended with aluminum powder, pressed into pellets, and then placed into targets that are inserted into

---

[18] Separated neptunium is a special nuclear material that, in specified forms and quantities, could be used to produce a nuclear explosive device.

a reactor to be irradiated in order to convert neptunium into Pu-238 isotope. Under the Supply Project, DOE officials plan to use two DOE research reactors—the High Flux Isotope Reactor (HFIR) at ORNL, which is currently used, and the Advanced Test Reactor (ATR) at INL, which is planned for use to convert the neptunium to Pu-238 in the future. Targets for ATR will be shipped from ORNL to INL for irradiation and then shipped back to ORNL for chemical processing, a necessary step to separate the newly produced Pu-238 isotope from unconverted neptunium and other byproducts from the process. Chemical processing also involves converting Pu-238 isotope into its oxide form, which is the form of Pu-238 used as fuel for RPS. Unconverted neptunium is recycled to be reused in target fabrication, and other waste materials are disposed.

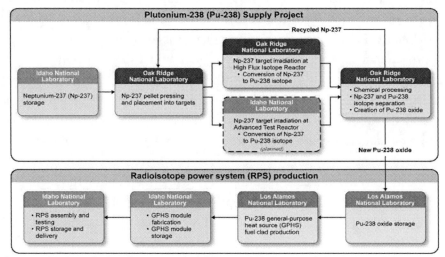

Source: GAO analysis of Department of Energy Documents. | GAO-17-673.

Figure 2. Overview of the Plutonium-238 (Pu-238) Supply Project and Radioisotope Power System (RPS) Production Process.

RPS Production across DOE's National Laboratories

RPS production occurs across three of DOE's national laboratories—ORNL, LANL, and INL. Until 2017, DOE officials managed RPS production across the national laboratories using a mission-specific approach. In March 2017, DOE officials, in consultation with NASA, moved from mission-specific production of GPHS fuel clads to a constant production rate. According to DOE officials, this change was made to provide stable staffing levels and maintain production capabilities across DOE's laboratories, among other things. Work at these laboratories requires specialized facilities, such as hot cells and glove boxes, and highly trained and qualified staff, because Pu-238 is highly radioactive.[19] These laboratories submit monthly reports to provide updates on RPS and Supply Project activities to DOE's Office of Nuclear Energy. Figure 2, above, provides an overview of DOE's RPS production process at ORNL, LANL, and INL. For example:

- ORNL is responsible for many of the processes related to DOE's efforts to reestablish a domestic supply of Pu-238. As such, Supply Project management is based at ORNL.[20] ORNL designs, tests, and qualifies neptunium targets for the Supply Project. Under the Supply Project, ORNL intends to ship newly produced Pu-238 to LANL for storage and use in new GPHS fuel clads.
- LANL maintains capability for Pu-238 processing and GPHS fuel clad production, among other mission support activities. This work is conducted at LANL's Plutonium Facility PF-4 and involves Pu-238 storage, chemical processing, analysis, fuel processing, and encapsulation of Pu-238 into GPHS fuel clads used in RPS.
- INL maintains capability for RPS assembly, testing, storage, and delivery of RPS for NASA. INL is also responsible for the transport

---

[19] According to DOE documents, a hot cell is a heavily shielded enclosure for handling and processing—by remote means or automatically—or storing highly radioactive materials. Remote operation refers to mechanical handling of irradiated nuclear materials by certain means, such as a robotic arm, to eliminate human contact with the materials. Gloveboxes are enclosures that enable operators to use their hands to manipulate hazardous materials through gloves without exposure to themselves or subsequent unfiltered release of the material to the environment.

[20] While ORNL manages the Supply Project, DOE's Technical Integration Office at INL aids in coordinating the Supply Project across the three laboratories.

equipment and logistics related to delivering RPS to NASA's Kennedy Space Center in Florida, as well as for supporting NASA at the launch site. Figure 3 shows an MMRTG being prepared for use in the Curiosity rover at Kennedy Space Center in 2011.

Source: National Aeronautics and Space Administration. | GAO-17-673.

Figure 3. Multi-Mission Radioisotope Thermoelectric Generator (MMRTG) for the Curiosity Rover at Kennedy Space Center.

NASA Funding for RPS Production

In 2011, NASA began fully funding DOE's Supply Project, and since 2014, NASA has been responsible for funding all aspects of RPS production operations and analysis to support launch safety and approval, according to NASA documents.[21] NASA funds DOE's efforts to build, test, and fuel RPS, as well as to update equipment and sustain staffing levels associated with

---

[21] Prior to 2014, DOE provided funding for infrastructure related to RPS production at DOE facilities, and NASA provided funding for mission-specific RPS production.

RPS production between missions. Specifically, NASA has provided, on average, approximately $50 million per year to support DOE's ongoing operations and maintenance of RPS production equipment. NASA provides additional funding related to specific missions that require RPS. According to NASA and DOE officials, DOE provides direct input during NASA's annual budgeting process, giving NASA officials information on DOE's capabilities and resource needs as they relate to the Supply Project and RPS production. See table 1 for NASA and DOE funding of RPS production activities, including the Supply Project.

NASA's and DOE's general RPS production roles and responsibilities are established in a memorandum of understanding agreed to in 1991 and revised in October 2016. The 2016 memorandum of understanding delineates the authorities and responsibilities of each agency related to, among other things, research, development, design, and production with respect to RPS. Under the framework of the memorandum of understanding, NASA and DOE have set up interagency agreements to establish high-level milestones and funding profiles for specific RPS and Pu-238 related work. For example, an April 2014 interagency agreement was issued that included NASA's request that DOE maintain the unique program, facility, and safety capabilities required to produce RPS and RPS components and provide related services in support of NASA missions. The agreement also specified that DOE will coordinate with NASA at least annually on the short- and long-range planning of resources, including possible allocations of hardware, facilities, staff, and Pu-238 to potential NASA missions. Another interagency agreement from January 2015 identified commitments for the planned Mars 2020 mission that, among other things, specified the number of RPS required and their expected power requirements and described expected coordination responsibilities. In addition, according to DOE officials, DOE and NASA have established periodic reporting requirements and management meetings to report progress and challenges related to RPS production and the Supply Project.

Table 1. Radioisotope Power System (RPS) Funds, Fiscal Years 2011 to 2017 (Dollars in thousands)

	Activity	2011	2012	2013	2014[a]	2015	2016	2017
Department of Energy (DOE)	DOE funds for RPS	46,906	64,524	60,707	$0	$0	$0	$0
National Aeronautics and Space Administration's (NASA) Planetary Science Division (PSD)	Plutonium-238 Supply Project	3,450	10,000	10,000	14,500	17,000	17,700	20,000
	Multi-Mission Radioisotope Thermoelectric Generator[b]	12,341	2,978	16	32	4,000	14,560	14,870
	Advanced Stirling Radioisotope Generator[c]	40,908	69,924	31,976	3,000	300	0	0
	Operations and maintenance of RPS production equipment	0	0	0	51,300	57,400	55,800	55,500
	PSD funds for RPS	*56,699*	*82,902*	*41,992*	*68,832*	*78,700*	*88,060*	*90,370*
	Total RPS funds	*103,605*	*147,426*	*102,699*	*68,832*	*78,700*	*88,060*	*90,370*

Source: NASA and DOE | GAO-17-673.

[a] Beginning in fiscal year 2014, funding operations and maintenance of RPS equipment came from NASA funding.

[b] Multi-Mission Radioisotope Thermoelectric Generator funding includes funding for NASA's Mars 2020 and Mars Space Laboratory missions.

[c] The development of the Advanced Stirling Radioisotope Generator was discontinued in fiscal year 2014 for several reasons, including costs.

NASA SELECTS RPS FOR MISSIONS BASED PRIMARILY ON THE AGENCY'S SCIENTIFIC OBJECTIVES, AND ADDITIONAL FACTORS COULD AFFECT DEMAND FOR RPS AND PU-238

NASA selects RPS to power missions based on the agency's scientific objectives and mission destinations. Multiple factors, including costs associated with RPS and missions, could affect demand for RPS and Pu238.

NASA Selects RPS to Power Missions Based on the Agency's Scientific Objectives and Mission Destinations, and Power Source Selections are Made Early in NASA's Mission Review Process

According to the NASA officials we interviewed, NASA selects RPS to power its missions based on the agency's scientific objectives and mission destinations. Generally, the need for RPS is apparent based on the mission's scientific objectives and destination, according to NASA officials we interviewed. For instance, an RPS is more likely to be needed for a mission to a distant planet or permanently shadowed crater where there is minimal sunlight.

According to NASA officials we interviewed, NASA prioritizes missions identified in the National Academy of Sciences' decadal survey report, which include missions that respond to NASA's scientific objectives and may require the use of RPS.[22] In addition, the National Aeronautics and Space Administration Transition Authorization Act of 2017 states that the NASA Administrator should set science priorities by following guidance provided in this decadal survey report.[23] The most recent decadal survey

---

[22] The National Academy of Sciences' decadal survey report, which represents the highest priority space exploration interests of the scientific community, presents a 10-year program of science and exploration with the potential to yield revolutionary new discoveries.

[23] Pub. L. No. 115-10, § 501(a)(2), 131 Stat. 18, 48.

report identified 16 potential missions, as shown in appendix II.[24] Ten of these potential missions were suggested as options for the decade from 2013 to 2022, and RPS was the recommended power source for 7 of the 10 missions. The decadal survey report also identified 6 missions for the decade beginning in 2022 and recommended using RPS as the power source for 3 of these missions.

NASA officials said that in some cases it is not immediately clear whether RPS or a different power source would be appropriate for a mission's destination—e.g., Jupiter and its moons and, to a lesser extent, Saturn. In such cases, an independent review team would look at aspects of a proposed mission that are unique or require further evaluation. For example, NASA used an independent review team to further evaluate whether the Europa Clipper mission could be successfully accomplished using solar power because typically a mission to one of Jupiter's moons would require the use of RPS. From March 2012 to August 2014, NASA officials examined and assessed the baseline power requirements for the Europa Clipper mission and worked with an independent review team to determine whether solar power was a feasible power option and would not compromise the mission's scientific objectives.[25] NASA officials we interviewed said they were surprised that solar power was determined to be feasible for the Europa Clipper mission.

After NASA chooses a mission, the power source for that mission is typically considered and selected many years before a mission is launched.[26] Generally, NASA conducts mission concept studies to demonstrate why using RPS or another power source is most appropriate to meet a given mission's scientific objectives.[27] Subsequently, the power source for a mission is also

---

[24] The most recent report was published in 2011, before the Supply Project was started, and stated that some of its recommended missions cannot be accomplished without new Pu-238 production.

[25] Specifically, the independent review team performed a solar feasibility review examining potential radiation exposure on solar cells that would mimic conditions near Jupiter and one of its moons, Europa.

[26] 26For example, NASA conducted its Mars 2020 mission concept review in August 2013, about 7 years prior to its anticipated launch date. RPS production for Mars 2020 began in 2015, 5 years before its launch.

[27] NASA officials said that mission concept studies are typically conducted for directed missions, and, for competed missions, it is the responsibility of the principal investigator to conduct the

considered early in NASA's formal mission review process—the lifecycle review process (see fig. 4). Specifically, NASA officials said that many aspects of a mission, including power sources, are considered during pre-Phase A, which is the first step of NASA's lifecycle review process.[28] At the conclusion of pre-Phase A, NASA's mission teams present the Mission Concept Review to a review board for consideration and to seek formal agency approval. A Mission Concept Review presentation outlines the preferred power source for a mission.[29]

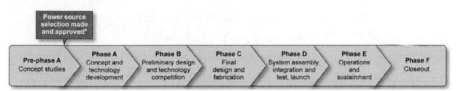

Source: GAO analysis of National Aeronautics and Space Administration guidance. | GAO-17-673.

[a] Competed missions do not go through pre-Phase A and Phase A of NASA's lifecycle review because reviews of competed missions are conducted through the competitive review process.

Figure 4. Power Source Selection in NASA's Lifecycle Review Process.

NASA officials we interviewed said that while power source decisions are reassessed throughout the lifecycle review process, once they are made, changes to mission power source decisions are rare. In one instance, NASA officials said that the Solar Probe mission originally called for the use of RPS, but a decision was made to choose another power source due to cost

---

necessary studies to decide upon a power source. In addition, NASA gives principal investigators an estimate for expected RPS costs, and the principal investigator is responsible for fitting that cost into a given mission's budget. As was the case for New Frontiers #3, if NASA does not make RPS available for a mission, principal investigators cannot propose missions that can only be accomplished with RPS.

[28] NASA officials said that competed missions (Discovery and New Frontiers) do not go through pre-Phase A and Phase A of NASA's lifecycle review because reviews of competed missions are conducted through the competitive review process. The principal investigator produces a mission concept study report equivalent to directed missions' Mission Concept Review. When NASA evaluates competed mission proposals, power sources such as RPS are not an explicitly defined selection criterion. After selection, competed missions are subject to the same lifecycle review process as directed missions.

[29] For instance, the Mars 2020 Mission Concept Review presentation included the proposal to use an MMRTG, similar to the Curiosity mission.

constraints and mission objective downsizing. More specifically, following a narrowing of the scope of the Solar Probe mission, RPS is no longer required, and the mission will instead rely on solar power.[30]

NASA officials said they also use review boards to provide independent assessments of mission planning at the end of each phase of the lifecycle review, including an assessment of power sources and margins.[31] NASA conducts these assessments at the conclusion of Phases B through E. These assessments address questions about the electrical power system for the mission and consider whether the selected power source can provide sufficient electrical power margins, based on the current spacecraft design, for example.

While NASA officials generally decide on a mission's power source at the end of pre-Phase A of its lifecycle review process, NASA officials emphasized that the official decision to use RPS is contingent on a final environmental impact statement and record of decision, in accordance with National Environmental Policy Act of 1969 requirements.[32] In addition to its lifecycle review process, NASA officials added that, consistent with the National Space Policy, RPS are used when they enable or significantly enhance a mission, such as when a power alternative, such as solar power, significantly compromises mission objectives. The National Space Policy specifically states that RPS shall be developed and used only when it

---

[30] According to NASA officials, the new mission, called the Parker Solar Probe, is scheduled for launch in 2018, is intended to provide new data on solar activity, and will aid NASA's ability to forecast major space-weather events that impact life on Earth.

[31] In this context, margin is the spare amount of mass or power allowed or given for contingencies or special situations.

[32] Under the National Environmental Policy Act of 1969, agencies evaluate the likely environmental effects of projects they are proposing using an environmental assessment or, if the projects likely would significantly affect the environment, a more detailed environmental impact statement. 42 U.S.C. § 4332(2)(C), (E). According to NASA officials, before the official record of decision for the act, power source considerations are notional, where using an RPS is considered only as a proposal. NASA officials added that an environmental impact statement is typically done to coincide with Phase A of NASA's lifecycle review process. NASA may be working almost exclusively on one potential power source for a mission, but other power options must remain in consideration until the record of decision for the act has been finalized, according to NASA officials.contingencies or special situations.

significantly enhances space exploration or operational capabilities.[33] Prior to the establishment of DOE's Supply Project in fiscal year 2011, mission selections were influenced by the limited amount of available Pu-238, NASA officials said. These same officials told us that missions are now selected independently from decisions about how they will be powered. However, projected availability of Pu-238 is factored into whether it is offered for a specific mission opportunity.

Multiple Factors Could Affect Demand for RPS and Pu238

Costs Associated with RPS and Missions Can Affect RPS Demand

NASA officials we interviewed said that the demand for RPS is driven by a mission's cost.[34] Based on NASA's expected funding levels, these officials said, NASA can support no more than one mission using RPS about every 4 years—or two to three missions per decade. NASA's current plans for solar system exploration—one RPS on the Mars 2020 mission and up to three RPS to support the selected New Frontiers #4 mission—is consistent with this mission frequency over the next decade. NASA officials said that they try to work within their budget to pursue and complete the right number of missions to meet the needs of the scientific community and to be consistent with the agency's scientific objectives. According to NASA officials we interviewed, when NASA selects a mission that requires RPS, the cost of the RPS must be supported by the mission's budget.

According to NASA officials, RPS have typically been used on Flagship missions that cost $2 billion or more. Flagship missions are generally very challenging, require more power for scientific instruments, and have larger budgets that can accommodate the costs of RPS. NASA estimates that a single RPS costs about $77 million, which would account for less than 5

---

[33] U.S. Office of Science and Technology Policy, *National Space Policy of the United States of America* (Washington, D.C.: June 28, 2010).

[34] NASA's PSD conducts missions using RPS. Approximately $1.6 billion of NASA's $19.3 billion overall appropriation was allocated to PSD in fiscal year 2016.

percent of a Flagship mission's overall cost.[35] Up to four RPS have been used in past missions; however, planned missions are projected to use one to three RPS per mission. Flagship mission planning and development occurs over multiple years, and the costs, including those related to RPS, are spread across that time frame. NASA officials we interviewed said that they are currently in the development phase for one Flagship mission—Mars 2020— and the formulation phase for a second Flagship mission—Europa Clipper. NASA officials said that regardless of the availability of RPS, NASA cannot support additional Flagship missions until after the expected launch of Mars 2020 in July 2020, given their expense relative to NASA's overall budget. Mars 2020 has an estimated total cost of $2.44 billion and will use one RPS.[36]

According to NASA officials we interviewed, New Frontiers missions, which are selected through a competitive process, may be good candidates to use RPS given the types of instruments needed to address the scientific questions these missions seek to answer. However, New Frontiers missions have a development cost cap of $850 million, which could restrict the use of RPS. A single RPS would represent about 9 percent of the development cost of a New Frontiers mission. NASA documents indicate that the principal investigator chosen to develop a New Frontiers mission is responsible for fitting the cost of the RPS into the mission's budget. In December 2016, NASA issued an announcement of opportunity for the New Frontiers #4 mission, offering up to three RPS.[37]

RPS have not been used for previously completed Discovery-class missions because these missions had smaller budgets—about $450 million—

---

[35] This estimate includes special services and other work associated with the launch of radioactive materials. However, this estimate does not include environmental reviews, which could cost $1 million or more to complete.

[36] NASA officials decided that another Flagship mission, the Europa Clipper mission that is expected to launch no earlier than May or June 2022, with a cost estimate range of $3.1 billion to $4 billion, will not use RPS.

[37] According to NASA documentation related to this announcement of opportunity, the use of one MMRTG—the current RPS design—would cost $77 million, two MMRTGs would cost $94 million, and three MMRTGs would cost $117 million. With an $850 million development cost cap, one MMRTG would account for 9 percent of the New Frontiers budget, two MMRTG would account for 11 percent, and three MMRTG would account for almost 14 percent of the mission's budget.

and shorter mission planning cycles—no more than 36 months— according to the National Academy of Sciences' most recent decadal survey report and NASA officials we interviewed. According to NASA officials, Discovery-class missions are small, relatively quick missions that typically rely on solar panels and not RPS as a power source. The cost of RPS would represent a large portion of a Discovery mission budget, according to NASA officials. A single RPS, at a cost of $77 million dollars, would represent more than 17 percent of a Discovery mission's $450 million development cap. Furthermore, the amount of time DOE takes to build an RPS generally exceeds Discovery-class mission development time frames of no more than 36 months and so it is unlikely that Discovery-class missions can use RPS unless limitations from cost and schedule can be reduced.

DOE's RPS Production Capability Can Limit RPS Use

In addition to budget factors, DOE's RPS production capability can limit NASA's ability to use RPS to power missions. According to DOE officials we interviewed, it can take up to 6 years to acquire, fuel, test, and deliver a new RPS for a NASA mission. According to both DOE and NASA officials we interviewed, DOE only has the capacity to produce three to four RPS at a time, given the current floor space dedicated to RPS development at INL and limits on staff exposure to radiation at LANL. NASA officials said that they would need to provide additional resources to scale up RPS production beyond its current levels. To accommodate DOE's current RPS production capability, NASA officials said they will not select two consecutive missions requiring RPS. For example, if NASA officials know that a Flagship mission will require RPS, they will not offer RPS for a competed mission that will launch around the same time. NASA officials we interviewed said that they provide DOE with a mission forecast, which provides prospective dates for missions offering RPS flight opportunities and gives DOE the associated power requirements for the missions.

Expected RPS and Solar Technological Advances Could Reduce NASA's Demand for RPS and Pu-238

NASA's Glenn Research Center is developing new RPS and solar power technology that may reduce NASA's demand for Pu-238 and thus RPS. NASA officials we interviewed said that they are working on technology advancements in order to preserve Pu-238, which they described as a scarce and expensive resource. Furthermore, the National Academy of Sciences' most recent decadal survey report advocates for RPS technology developments, stating that it is imperative that NASA invest in technology for likely future missions, with the goals of reducing costs and improving scientific capability and reliability.

According to a report to Congress, in 2010, NASA anticipated using a dynamic RPS design, such as the Advanced Stirling Radioisotope Generator (ASRG), for future missions.[38] This design, according to NASA documentation, would have provided a more efficient power system than the MMRTG RPS design currently in use.[39] DOE's initial Pu-238 production goal of up to 5 kg per year from 2001 was reduced to 1.5 kg based on the assumption that a dynamic system like the ASRG would be available. The ASRG was projected to be more than four times as efficient as the MMRTG and would have required just one-fourth as much Pu-238. However, according to NASA officials, in 2013, the agency ceased ASRG development due to the cost to complete the project under reduced PSD funding (see table 1 for NASA's prior funding of RPS production activities, including ASRG). Although the 1.5 kg-per-year Pu238 production goal was based on the improved efficiencies of the ASRG and a reduction in the amount of Pu-238 needed for missions, NASA officials said that the projected production from the Supply Project will nonetheless meet their demands using the MMRTG, based on PSD's current mission frequency. Furthermore, NASA's fiscal year 2018 budget estimate indicated that NASA

---

[38] U.S. Department of Energy, Start-up Plan for Plutonium-238 Production for Radioisotope Power Systems: Report to Congress (Washington, D.C.: June 2010).

[39] The National Academy of Sciences' decadal survey report from 2011 advocated for the continued development of an ASRG, a dynamic RPS design that involves moving parts, as opposed to the MMRTG, which is a static system with no moving parts. The ASRG design is purported to have more efficient power conversion than the MMRTG currently in use.

will again invest in dynamic power conversion systems. Specifically, NASA's Glenn Research Center officials said that current work to advance RPS technology would reduce the amount of Pu-238 needed. Officials told us that they plan to invest $8 million in dynamic RPS technology beginning in fiscal year 2018 and increase funding to $10 million by fiscal year 2022.

In addition to dynamic RPS design, NASA's Glenn Research Center officials said that they are researching advancements in thermoelectric technologies that could also increase RPS efficiency. According to these officials, NASA is currently funding this effort in fiscal year 2017 at $8 million and plans to increase funding to $9 million per year by fiscal year 2022.[40] NASA officials said that new thermoelectric materials can lead to improved power conversion efficiency and reduce the degradation of thermoelectric couples over time.[41] Specifically, NASA officials said NASA plans to update the current MMRTG design using a new thermoelectric couple material called skutterudite. NASA officials said that this enhanced MMRTG (eMMRTG) would have a 25 percent beginning-of-life efficiency improvement and a 50 percent increase of end-of-life power output. The power output for the MMRTG is about 120 watts, while the eMMRTG is expected to produce from 120 to 160 watts. According to NASA officials, for a mission requiring 300 watts of power, only two eMMRTGs would be needed instead of three MMRTGs. According to NASA officials, using one less RPS unit would save about 3.5 kg of Pu-238 isotope and reduce the mission's overall mass by 45 kg. A NASA official said that the thermoelectric couple technology for the eMMRTG is currently assessed to be at a technical readiness level 3 and will be assessed for technical readiness

---

[40] A portion of the funding—$5 million—will be used to look at the chemistry of thermoelectric elements, and once these are better understood, NASA intends to develop them into thermoelectric couples to be tested through NASA's advanced technology maturation project, according to NASA officials.

[41] Thermoelectric couples are pairs of electrically conductive materials joined in a closed circuit with each side of the pair kept at different temperatures. The thermoelectric couples in the MMRTG use the heat from the decay of Pu-238 to heat the hot side, and the cold of space or planetary atmosphere on the cold side of each pair. However, the MMRTG experiences a power drop-off over time due to the natural decay of Pu-238 as well as the decreased performance of the thermoelectric elements due to high heat deterioration and sublimation of materials.

level 4 by the end of 2017.[42] NASA will determine at a later date if the technology is sufficiently advanced to use the new eMMRTG for the New Frontiers #4 mission.

In addition, NASA officials we interviewed said that they are investigating the next generation of RPS beyond the MMRTG and eMMRTG designs. For example, these officials said that they are reviewing a modular RPS device. A modular RPS would provide a smaller power output, but the modular units could be bundled together to provide a specific power output that meets mission requirements. This would allow NASA to use smaller power increments for missions, thus requiring less Pu-238 to provide that power. More specifically, NASA officials said that this type of power system could have a reduced mass and use a more precise amount of Pu-238 than is required for a mission, depending on the mission's power needs. According to NASA officials, the next generation RPS would ideally achieve up to 600 watts if using two next-generation systems for New Frontiers missions and about 300 watts for Discovery- class missions. Although NASA officials said that they anticipate the Supply Project's Pu-238 production rate goals to be sufficient to meet their mission demands using the existing MMRTG, if advances in conversion efficiency of future RPS designs come to fruition, NASA's future demand for Pu-238 may decrease. Table 2 compares the current MMRTG with other potential RPS technologies.

According to NASA officials, advances in solar technology could increase the distance from the sun at which missions can operate, making it possible for missions that would otherwise require RPS to be powered, instead, by solar panels. NASA officials said they are working with contractors to focus on overcoming two of the main challenges with using solar panels that are specific to deep space missions—low-intensity, low-

---

[42] Technical readiness levels are a scale used to measure the maturity of a given technology. Technology readiness level 3 represents analytical studies and demonstration of nonscale pieces of a subsystem in the laboratory environment. Technology readiness level 4 is representative of a technically feasible approach, though not fully functional. It has been validated in the laboratory environment. See GAO, *Technical Readiness Assessment Guide: Best Practices for Evaluating the Readiness of Technology for Use in Acquisition Programs and Projects.* GAO-16-410G (Washington, D.C.: August 2016).

temperature conditions and high-radiation environments. These officials said that there have been advances in solar technology regarding how much energy can be captured and converted into electricity, and these advances could help address low levels of light intensity. NASA officials said that because low temperatures can degrade some solar cells, the agency is pursuing solar cells that are better suited for cold environments. NASA officials said that advances in solar power technology have realistically expanded the ability to use solar power for missions for which it would not have been considered before. For example, these officials noted that NASA's current Juno mission and its planned Europa Clipper mission—both with destinations being at Jupiter and its moons and which would typically have used RPS—have demonstrated that solar power is a viable power option.[43]

Table 2. Radioisotope Power Systems for Space Exploration

	Current	Potential Performance	
System	Multi-Mission Radioisotope Thermoelectric Generator (MMRTG)	Enhanced Multi-Mission Radioisotope Thermoelectric Generator (eMMRTG)[a]	Advanced Stirling Radioisotope Generator (ASRG)[b]
Power	120 Watts (W)	120-160 W	120-160 W
Lifetime	17 years	17 years	17 years
Efficiency (in percent)	6	8-12	12-30
Pu-238 Oxide	4.8 kilograms (kg)	4.8 kg	1.2 kg
General purpose heat source fuel clads	32	32	8

Source: GAO analysis of National Aeronautics and Space Administration documents. | GAO-17-673.

[a] The eMMRTG is still in development. The eMMRTG would use the same platform and components as the MMRTG with the exception of the materials used in thermoelectric couples.

[b] In 2013, NASA stopped funding ASRG because of costs, among other things. However, funding for dynamic power conversion development is included in NASA's fiscal year 2018 budget estimate.

---

[43] According to NASA officials, advances in solar technology have been able to concentrate light intensity at Jupiter to near Earth levels. Light intensity at Jupiter is 1/50th of the light intensity at Earth.

NASA Does Not Anticipate RPS and Pu-238 Demand from Other Users

NASA does not anticipate other potential users to affect demand for RPS or Pu-238. According to DOE planning documents and NASA officials, expected RPS and Pu-238 production from the Supply Project is intended to only meet PSD's demand. The Supply Project goal of 1.5 kg of Pu-238 per year was established to support two to three PSD missions using RPS each decade. NASA officials said that they did not account for demand from other potential users within NASA or national security and commercial sectors when establishing production goals for Pu-238. While NASA does not expect other users to affect demand for Pu-238, there are other potential uses for Pu-238 outside PSD, including precursor missions in support of human exploration missions or to provide RPS for national security uses.

Specifically, NASA Human Exploration and Operations (HEO) officials we interviewed said that RPS could be used to power precursor missions to Mars in advance of human exploration missions. As of now, HEO officials have identified only one potential precursor mission—a prospecting rover that would test equipment needed to convert atmospheric gas on Mars into fuel—but the baseline power option for this mission would likely be solar. In addition, NASA officials said that they do not anticipate using RPS for primary power on missions involving manned surface rovers or human inhabitants because, even with extreme production projections, there cannot be enough Pu-238 to provide sufficient power. According to NASA officials, however, they are in the early stages of planning for potential future human exploration and precursor missions, and through this planning process, they will determine whether RPS have a potential role in these missions.[44] Because such potential HEO missions are not factored into the current production goal of 1.5 kg of Pu-238 per year, if HEO officials determine that RPS are needed, such as for providing heat or auxiliary

---

[44] According to PSD officials, NASA's early demand estimate for 5 kg of Pu-238 per year included both PSD and HEO mission needs; however, since then, HEO has moved away from the use of RPS. According to HEO officials, RPS do not provide sufficient power to support human exploration missions.

power, HEO would need to coordinate with the PSD to make mission prioritization decisions.

NASA officials said they have been approached in the past about RPS availability for privately funded space missions, though not in several years. If commercial entities were to require RPS for their missions, they would need to develop a partnership with NASA and DOE to acquire RPS. Additionally, despite being approached in the past, NASA officials said they currently have no plans to offer RPS for privately funded space missions. Similarly, DOE officials view the use of Pu-238 in these contexts as unlikely. As a result, NASA officials did not account for this potential use of RPS when establishing production goals for the Supply Project. These officials said that if any significant quantity of Pu-238 was to be offered to private industry, NASA planetary exploration would be affected.

While DOE maintains a separate inventory of Pu-238 for national security purposes, if that inventory degrades below usable levels or more quantity is needed, according to NASA and DOE officials, NASA's expected Pu238 from the Supply Project could be reallocated for national security purposes. NASA officials said that DOE has the authority to govern the allocations of the Pu-238 inventory, and that any reallocation for national security purposes would be addressed by the executive branch and its agencies. In such circumstances, NASA officials said that they would work with the administration to address those priorities.

DOE HAS MADE PROGRESS REESTABLISHING PU238 PRODUCTION BUT FACES CHALLENGES THAT COULD AFFECT ITS ABILITY TO MEET RPS AND PU-238 DEMAND

DOE has made progress reestablishing Pu-238 production to meet NASA's RPS demand, is in position to support NASA's current plans for solar system exploration, and anticipates being able to support two to three RPS-powered missions per decade using new Pu-238 expected from the

Supply Project. However, DOE faces challenges with key aspects of RPS and Pu-238 production that could put production goals at risk.

DOE Has Made Progress Reestablishing Pu-238 Production to Meet NASA's RPS Demand

DOE has made progress reestablishing Pu-238 production to meet NASA's future demand for Pu-238 to fuel RPS. DOE officials said that they will be able to support NASA's current plans for solar system exploration and expect to be able to support two to three RPS-powered missions per decade for the foreseeable future using new Pu-238 expected from the Supply Project. A selected chronology of key planned DOE RPS and Pu-238 production activities, with NASA's mission-related activities, are shown in figure 5.

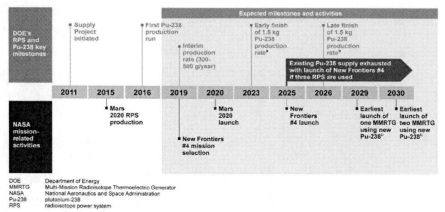

DOE　Department of Energy
MMRTG　Multi-Mission Radioisotope Thermoelectric Generator
NASA　National Aeronautics and Space Administration
Pu-238　plutonium-238
RPS　radioisotope power system

Source: GAO analysis of DOE and NASA information. | GAO-17-673.

[a] DOE has established a date range of 2023 to 2026 to achieve full production of 1.5 kg of Pu-238 per year; however, as of June 2017, DOE officials said they expect to reach full production no earlier than 2025.

[b] Earliest launch dates assume delivery of 1.5 kg of Pu-238 by the start of fiscal year 2026.

Figure 5. Selected Chronology of Department of Energy and National Aeronautics and Space Administration Radioisotope Power Systems and Plutonium-238 Production Activities.

A critical component of meeting NASA's expected RPS-related mission requirements is the production of new Pu-238. NASA officials said that they are confident that DOE will meet its Pu-238 production goals. DOE demonstrated a proof of concept for new Pu-238 production and has made approximately 100 grams of new Pu-238 isotope under its Supply Project since its inception in 2011. As a result of the newly produced Pu238 isotope quality, it can be blended with Pu-238 that does not meet NASA's spaceflight power specifications in order to extend the usefulness of NASA's existing Pu-238 supply. Both DOE and NASA officials said that they expect to develop and finalize a blending strategy after the quality of Pu-238 produced under the Supply Project is fully assessed. Given DOE's current Supply Project and RPS production schedule, and NASA's space exploration plans, the existing Pu-238 supply will be exhausted by 2025 if four total RPS are used for NASA's Mars 2020 and New Frontiers #4 missions. Furthermore, until DOE and NASA develop and finalize a blending strategy, the earliest RPS-powered mission using new Pu-238 would be in 2029, assuming delivery of 1.5 kg of Pu-238 to LANL for GPHS fuel clad production at the start of fiscal year 2026, as shown in figure 4, above. DOE's progress in meeting NASA's future Pu-238 and RPS demand are illustrated in the following examples.

Using Existing Facilities and Automating Supply Project Steps at ORNL

For the Supply Project, as well as for RPS production, DOE officials said that they are primarily leveraging existing facilities, equipment, and previously proven processes. DOE officials said that it would be prohibitively expensive to build new facilities for the Supply Project's chemical processing work. For example, DOE officials are using hot cells at ORNL that are also used for other isotope work.[45] In addition, according to DOE officials we interviewed, new equipment is being acquired at ORNL to modernize and automate some steps of the Supply Project.[46] For example,

---

[45] Existing hot cells at ORNL are used for other isotope production work, such as californium.
[46] According to DOE documentation, facility and infrastructure improvements at ORNL for the first segment of the Supply Project were estimated to cost $450,000. Beginning at the start of

DOE has acquired equipment to automate neptunium target production, which DOE officials said they expect will increase Supply Project efficiency as well as reduce staff radiation exposure. Preparing and pressing neptunium into pellets for use in targets is currently a manual process done in gloveboxes, with a press that was formerly used for iridium processing and that was built in the 1920s. DOE officials said that they are able to make about 40 neptunium targets per year, equivalent to pressing one neptunium pellet about every 35 minutes. To achieve DOE's interim goal of 300 to 500 grams of new Pu-238 per year by 2019 and to build up an inventory of targets for full production, DOE will need to make about 128 targets per year, or one neptunium pellet about every 6 minutes. DOE officials said that the automated system will help provide the necessary throughput to meet interim production goals. Without automation, DOE officials determined that they could produce from 120 to 150 grams of new Pu-238 per year. DOE officials said that they intend to have neptunium pellet automation implemented in 2017. The officials added that neptunium pellet automation is expected to increase efficiencies with respect to meeting documentation and quality control requirements prior to target irradiation in a reactor.

Upgrading RPS Fuel Production Equipment at LANL

DOE officials said that investments and equipment upgrades are also being undertaken at LANL in support of RPS production. Specifically, LANL officials said that from 2010 through 2012, they carried out a comprehensive programmatic equipment plan to maintain and upgrade equipment related to RPS production work. The plan is now being updated since major work has been completed, according to LANL officials. As part of this plan, LANL officials said they are undertaking efforts to revitalize and extend the life of gloveboxes and other equipment at PF-4 in support of Pu-238 activities. For instance, LANL monthly reports indicate that windows on a fuel production line have been changed out—a necessary improvement because of degradation caused by oxidation in the processing environment. According to LANL reports and officials, they are installing a

---

this segment in October 2015, through the planned end of the Supply Project in 2026, these improvements are estimated to cost a total of $1.74 million.

new Pu-238 fuel pellet hot press that will be used after the current production run of Pu-238 fuel clads for NASA's Mars 2020 mission is complete. The Pu-238 fuel pellet hot press currently in use is about 50 years old and needs frequent maintenance in order to keep it operational. The hot press currently in use will serve as a backup once installation of the new press is completed in 2019.

Testing and Fabricating RPS at INL

DOE officials told us that their main objective for RPS production at INL is to maintain core capabilities to fabricate RPS units. DOE officials said they maintain 30 pieces of equipment, totaling about $30 million in value, that are replaced or refurbished on a rolling basis. For example, INL maintains equipment to support RPS acceptance tests—vibration testing, mass properties testing, magnetics testing, and thermal vacuum testing— to ensure that RPS are suitable for space flight.[47] INL officials said that they do not have backups for most of the equipment so it is imperative that they enhance the equipment's reliability through repair or replacement. INL funding data show that INL sets aside about 5 percent of its annual operations funding provided by NASA for RPS production work to update equipment at INL. These officials said that over the last 12 years they have nearly completed updates to existing equipment related to RPS production.

Stabilizing Staff Levels with a Constant GPHS Production Rate

DOE officials we interviewed said that staff retention between missions has been a challenge and that hiring and training new staff to support new missions can take up to 2 years because of the highly technical nature of the job. According to DOE officials and documents, under the previous management approach, DOE retained some staff to support ongoing Pu-238 and RPS work at INL, ORNL, and LANL but would need to hire more staff to support mission-specific work. Specifically, according to INL documentation, there are 40 staff members who support ongoing maintenance and operation of RPS equipment during off-peak times. INL

---

[47] Vibration testing, for example, is conducted to ensure that the RPS can withstand forces during a launch that could destroy the RPS.

officials said that they would need to hire 15 to 20 people, to reach 55 to 60 staff members, to support mission-related activities under the mission-specific production rate at their laboratory.

DOE officials said that the constant GPHS fuel clad production rate will provide stable staffing levels, removing the need to ramp up hiring and training of qualified staff for mission-specific RPS production. However, according to DOE officials, there are limits on how much DOE can increase GPHS fuel clad production. DOE may only be able to produce 10 to 15 GPHS fuel clads per year—consistent with DOE's current production rate when preparing for a NASA mission and a sufficient quantity to meet currently planned NASA missions. According to DOE officials, if NASA were to require additional GPHS fuel clads for RPS—for example, from 15 to 20 per year—DOE would face challenges in increasing production due to risks to staff from potential radiation exposure at LANL. Additionally, LANL officials said that if NASA requires an increase in the GPHS fuel clad production rate, the current Pu-238 supply would be depleted by 2022, before the Supply Project is fully operational.[48] DOE has instructed its facilities to develop plans to carry out constant rate production and to begin transitioning to the constant production rate model by June 2017.[49] DOE anticipates that NASA will sign new interagency agreements and will provide funding to implement constant GPHS fuel clad production in fiscal year 2017.

DOE Faces Challenges with Key Aspects of Pu-238 and RPS Production

DOE officials from INL, LANL, and ORNL identified several challenges that need to be overcome for DOE to be able to meet its projected Supply Project goal of 1.5 kg per year of Pu-238 by 2026, at the latest.

---

[48] The Supply Project is expected to be fully operational in 2023 at the earliest and in 2026 at the latest.

[49] The Office of Nuclear Energy issued a memo on March 7, 2017, to its facilities, INL, LANL, and ORNL, to develop a plan within 45 days and to begin operating under the constant production rate model within 90 days.

Otherwise, DOE may not able to reach its full production rate or may need to delay its delivery of 1.5 kg of Pu-238 per year. Specifically, these officials identified perfecting and scaling up chemical processing, the availability of reactors, and the qualification of targets for irradiation as challenges that need to be overcome.

Perfecting and Scaling Up Chemical Processing

DOE faces challenges perfecting and scaling up chemical processing that, if not overcome, could result in delays in producing Pu-238 to support future NASA missions. According to DOE officials, DOE is still in the experimental stage and has not perfected the chemical processing required to extract new Pu-238 isotope from the irradiated targets, which creates a bottleneck in the Supply Project and puts production goals at risk. ORNL monthly reports and DOE officials state they are continuing to develop the process to chemically separate newly produced Pu-238 isotope from unconverted neptunium and other materials resulting from irradiation in the reactor. According to officials, DOE is exploring chemical processing methods to effectively recycle neptunium for reuse in the process, convert Pu-238 isotope into its oxide form for use in RPS, and improve methods to reduce radioactive liquid waste.[50]

According to ORNL monthly reports and officials we interviewed, two 5- gram samples of Pu-238 taken from the first 50 grams of new Pu-238 from the Supply Project were shipped to LANL and tested in January and February 2016. These initial samples, however, did not meet space flight specifications.[51] Office of Nuclear Energy officials said that samples from a second 50-gram batch were tested at both ORNL and LANL in October 2016 and met flight specifications. According to DOE officials, chemical processing cannot be done in a linear fashion—that is, the chemicals used in the process are not all increased in the same quantities when making

---

[50] Office of Nuclear Energy officials said that improving the neptunium recycling process associated with chemical processing is not a necessary step to meet its production goals.

[51] LANL officials said that a specific chemical composition for Pu-238 is required in order to meet flight specifications. For example, the irradiation of neptunium-237 to create Pu-238 creates unwanted byproducts, such as thorium, that must be removed during the chemical processing phase, according to LANL officials and an ORNL report.

increasing amounts of new Pu-238. As a result, DOE officials said that they need to develop a chemical processing model that will help identify bottlenecks and provide information to help DOE officials improve the Pu-238 production process. However, ORNL officials said that chemical processing needs to be scaled up to meet interim and final production goals for Pu-238. For example, ORNL officials said they have sufficient staff to process one batch of irradiated targets at a time to extract new Pu-238 isotope, though multiple batches would need to be run simultaneously to achieve production goals.

According to ORNL documents and officials, to reduce the risk of failing to achieve Pu-238 production goals, additional resources for staff and equipment are needed to increase chemical processing operations. Specifically, to achieve interim production of 300 to 500 grams of Pu-238 per year by 2019, ORNL officials said they would require additional funding of $3 million to 4 million per year for staff and other process improvements. In addition, to reach full-scale operation of 1.5 kg of Pu238 per year by the end of fiscal year 2023, even more staff and funding are needed. However, these officials did not specify how many new hires would be needed to meet interim and full-scale operations. ORNL officials said that because of the highly technical nature of the job, it can take 2 years to train staff to use specialized equipment. Furthermore, DOE officials we interviewed at ORNL said that the laboratory needs infrastructure improvements for chemical processing, including additional storage tanks, transfer lines, and glove boxes, in order to increase Pu238 production beyond current sample-sized levels. DOE's ability to meet its production goal and support future NASA missions is at risk if the chemical processing steps in the Pu-238 production process are not improved and scaled up.

Availability of Reactor Positions for Target Irradiation

Officials we interviewed at INL and ORNL also said that achieving 1.5 kg of Pu-238 per year is contingent on the availability of positions within both the High Flux Isotope Reactor (HFIR) and the Advanced Test Reactor

(ATR) to irradiate neptunium targets for conversion to Pu-238 isotope.[52] DOE officials stated that ATR must be used, as projected in DOE's initial plans, to reach the production goal of 1.5 kg of Pu-238 per year.

However, ATR has not been qualified for Supply Project work because NASA has decided to wait until the reactor returns to service in 2021, after a yearlong scheduled maintenance shutdown beginning in 2020.[53] Until ATR is qualified, it is not available for new Pu-238 production. DOE officials responsible for the Supply Project have utilized HFIR at ORNL to irradiate targets, but the forecasted Pu-238 production from this reactor is anticipated to be less than initially planned because target positions within HFIR are limited so as to not interfere with other reactor activities. DOE's 2013 analysis of Pu-238 production projected that HFIR could produce 2 kg of Pu-238 per year, but according to DOE officials, to avoid interference with HFIR's other missions, they subsequently determined that HFIR would produce approximately 600 grams of Pu-238 isotope with the current target design. According to DOE officials, they plan to use positions within ATR to achieve full production goals.

In addition, deferring ATR's use could put achieving production goals at risk because accessing positions within ATR is highly competitive. DOE officials at INL have said that ATR's availability for the Supply Project may be limited due to competition from other users, such as the U.S. Navy, universities, medical researchers, and other DOE programs. Specifically, DOE officials at INL said that the positions most useful for irradiating neptunium targets to create Pu-238 isotope are over-utilized, and of the nine available positions for target irradiation, six are dedicated for Navy use.[54] Despite this, Office of Nuclear Energy officials said that they do not foresee an impact on meeting Pu-238 production goals from deferring the use of

---

[52] Positions are locations within the reactors where targets are bundled and placed for the irradiation process. Only certain positions are suitable for Pu-238 production. According to DOE documentation, HFIR has 22 positions within the reactor, of which 20 are suitable for Pu-238 isotope production. According to INL documentation and officials, ATR has 75 positions within the reactor, of which 9 are suitable for Pu-238 isotope production.

[53] DOE officials said that reactors require periodic shutdowns to replace necessary shielding for safety purposes, among other things.

[54] DOE officials at INL said that the U.S. Navy uses ATR for work related to extending the life of fuel on naval ships.

ATR. These officials said that DOE conducted a trade study to identify positions within ATR that are not currently being used by the Navy or other ATR users, in order to determine whether they are suitable for Pu-238 isotope production. According to the study, several such positions could be suitable for Pu-238 production, but DOE officials said that additional verification testing would be conducted in fiscal year 2017 to confirm such findings. However, DOE officials said that if positions in ATR are not available for Pu-238 isotope production, they do not have a plan to address this longer-term challenge and will be unable to meet full Pu238 production goals.

Target Design and Qualification

While DOE officials said that they expect to meet Pu-238 production goals with the existing target design, they have not qualified the existing target for use in ATR, which is needed to reach 1.5 kg of Pu-238 per year. In addition, DOE is pursuing a new target design with a higher expected Pu-238 yield. However, the new target design has not been developed and qualified for either HFIR or ATR. Furthermore, because the target has not been fully developed, its expected yield—double the amount of Pu-238 as the current target in certain reactor positions—remains theoretical. According to Office of Nuclear Energy officials, DOE and NASA have agreed to complete research and development efforts to support a decision on the new neptunium target by 2019 that can be used in both HFIR and ATR. ORNL documentation from July 2016 indicated that the laboratory's initial task is to test and assess the new target design on a small scale. According to DOE officials at ORNL, upon successful small-scale testing, they will scale up a prototype target for further testing.

DOE officials said that the higher yield from the new target design could allow DOE to meet or exceed interim and final production time frames and goals because they were calculated using the current target design as the baseline. Officials from the Office of Nuclear Energy said that the established Pu-238 production goals used the existing neptunium target design as their baseline and that any increased Pu-238 production expected with the new target design represents an opportunity. However, officials we interviewed at ORNL and INL said that the anticipated production from the new neptunium target design may be necessary in order to mitigate the effects of other challenges, such as limited reactor space, as discussed above. Office of Nuclear Energy officials said that the existing Pu-238 inventory and planned NASA missions will allow them time to finalize the target design and begin using ATR to provide irradiation capability starting in 2021. According to DOE estimates, it would take from 1 to 2 years to begin irradiating targets at ATR for new Pu-238 isotope production.

Prioritizing Pu-238 for NASA at Plutonium Facility 4

Another challenge DOE must overcome to achieve its RPS and Pu-238 production goals relates to potential competition for space between Pu238 and nuclear weapons pit production activities at LANL, which could significantly affect a key step in the RPS production process within the next decade, and thus delay the delivery of RPS for NASA missions. Specifically, LANL officials said that Pu-238 work must compete with other priorities for facility space at LANL's Plutonium Facility PF-4. As we stated in an August 2016 report, the National Nuclear Security Administration—a semi-autonomous agency within DOE that is responsible for the management and security of the nation's nuclear weapons, nuclear nonproliferation, and naval reactor programs—through its plutonium strategy is focused primarily on the fabrication of plutonium pits for nuclear weapons and has not coordinated with the Pu-238 program about potential modifications planned for PF-4.[55] A May 2015 Congressional Research Service report suggested that moving Pu-238 operations outside of the current LANL PF-4 facility could free up floor space for pit production operations, and also reduce the amount of nuclear material at risk in the facility due to the very high radioactivity of Pu-238.[56] However, LANL officials told us that such a move would be unlikely because of cleanup costs and difficulties in transporting contaminated equipment. In addition, DOE's Office of Nuclear Energy

---

[55] GAO, DOE Project Management: NNSA Needs to Clarify Requirements for Its Plutonium Analysis Project at Los Alamos, GAO-16-585 (Washington, D.C.: Aug. 9, 2016).
[56] Congressional Research Service, Nuclear Weapon "Pit" Production: Options to Help Meet a Congressional Requirement, R44033 (Washington, D.C.: May 14, 2015).

officials said that they do not anticipate pit production to impact RPS production. However, LANL officials told us that they have begun preliminary discussions on this issue but are awaiting the results of a National Nuclear Security Administration plutonium strategy study that is focused on nuclear weapons needs. DOE officials added that they have discussed moving Pu-238 capabilities out of PF-4, where they would seek to move the capability to another DOE site. According to DOE officials, a move of this nature would likely be an expensive, disruptive, and time-consuming effort.

Challenges May Put RPS and Pu-238 Production Goals at Risk

The challenges noted above—chemical processing, availability of reactor positions, target design and qualification, and prioritizing Pu-238 at PF4—may put DOE's RPS and Pu-238 production goals at risk, in part because of the short-term and incremental segmented management approach DOE has used to manage the Supply Project through early 2017. In this approach, DOE established short-term segments of Supply Project work to be connected to time frames over which DOE could more reliably predict funding from NASA. While DOE officials have known the general tasks required to achieve full production goals for the Supply Project, they have been relying on the developments of each segment to inform specific tasks and activities of subsequent segments. For example, up until March 2017, when DOE changed its management approach for the Supply Project, if challenges, such as delays in scaling up chemical processing, were to be realized, DOE officials said that they would have shifted Supply Project activities from one segment to the next. DOE's own work, however, demonstrates the risks in this approach. For example, an October 2016 DOE independent cost estimate for another DOE program showed that deferring activities to later stages in that program may result in program delays or, if delays are undesirable, future activities may be compressed and would likely be less technically mature. Delays in the Supply Project, or the use of less technically mature processes, could put DOE's ability to meet NASA's Pu-

238 demands at risk. DOE officials said in June 2017 that the Supply Project's early completion of full Pu-238 production was initially planned for 2023. However, DOE officials said they now expect to reach full production no earlier than 2025 with a late completion date remaining in 2026.

DOE officials, acknowledging that some long-term challenges to the Supply Project exist, said that they recognized that the segmented management approach was not well suited to dealing with those challenges. As previously noted, in March 2017, DOE officials said that the change to a constant GPHS production rate approach was expected to help provide funding flexibility and stabilize RPS production staffing levels between NASA missions. In June 2017, DOE officials said that the constant GPHS production rate approach would also address other identified challenges associated with RPS production and the Supply Project and decided to discontinue the segmented management approach. According to DOE officials, the constant GPHS production rate approach is designed to ensure stable and predictable funding levels and allow DOE more flexibility in how these funds are to be used. This approach, according to DOE officials, will allow the agency to address high-priority challenges by providing flexibility to use NASA funding throughout the RPS supply chain, including for the Supply Project, and will better enable them to sequence work. For example, DOE officials described how a constant GPHS production rate approach allowed DOE to prioritize funding for shipping containers to transport Pu-238 between laboratories.[57] DOE officials did not describe, however, how this new approach would help address some of the other longer-term challenges identified, such as scaling up and perfecting chemical processing. Furthermore, DOE officials provided basic details about the new approach as well as a memorandum dated March 7, 2017, with deliverables that DOE states that the new approach will achieve. However, DOE does not yet have an implementation plan under the new constant GPHS production rate

---

[57] For example, in response to Congress' proposed fiscal year 2017 omnibus appropriation bill, DOE recommended accelerating the procurement of equipment necessary to increase Pu-238 load-out capability at ORNL—including the development, design, and procurement of inner and outer shipping containers—rather than conducting this work in fiscal year 2019, according to a DOE document.

approach with milestones and interim steps for the Supply Project or for RPS production that can be used to show progress toward implementing efforts, show how risk is being addressed or mitigated, or make adjustments to those efforts when necessary. Our body of work has shown that without defined tasks and milestones, it is difficult for an agency to set priorities, use resources efficiently, measure progress, and provide management a means to monitor this progress.[58] By developing an implementation plan with milestones and interim steps for the department's management approach for Pu-238 and RPS production, DOE can show progress toward implementation, show how risk is being addressed or mitigated, or make adjustments to its efforts when necessary.

In addition, DOE's new approach still does not improve the agency's ability to assess the long-term effects of the challenges associated with Pu-238 and RPS production, such as chemical processing, availability of reactor positions, target design and qualification, and prioritizing Pu-238 at PF-4. DOE's previous segmented management approach did not require that DOE officials assess the potential long-term effects of these challenges, and it allowed them to defer addressing challenges to later segments of the Supply Project. The new approach also does not allow for DOE to adequately communicate long-term challenges to NASA, and it is unclear how DOE will use this approach to communicate these challenges. According to DOE officials we interviewed, DOE and NASA are to develop reports and hold management meetings to communicate identified challenges. In addition, *Standards for Internal Control in the Federal Government* states that agency management should use quality information to achieve the entity's objectives and communicate quality information externally through reporting lines so that external parties can help the entity achieve its objectives and address related risks.[59] DOE officials said they believe the new constant GPHS production rate approach will aid DOE in identifying

[58] GAO, Defense Health Care Reform: Actions Needed to Help Ensure Defense Health Agency Maintains Implementation Progress, GAO-15-759 (Washington, D.C.: Sept. 10, 2015), and Biobased Products: Improved USDA Management Would Help Agencies Comply with Farm Bill Purchasing Requirements, GAO-04-437 (Washington, D.C.: Apr. 7, 2004).

[59] GAO, Standards for Internal Control in the Federal Government, GAO-14-704G (Washington, D.C.: September 2014).

and addressing challenges associated with RPS production and the Supply Project, as well as communicating such challenges to NASA. DOE officials we interviewed, however, provided few details on how they would address identified challenges. Without adequately assessing the long-term effects of existing challenges identified by DOE officials, DOE cannot develop the quality information it needs to understand the effects of these challenges, and this may affect its ability to achieve its Pu-238 production goal or to communicate that information to NASA.

Moreover, because DOE does not maintain a comprehensive risk tracking system for RPS production and, instead, defers to its individual laboratories to track and manage their specific risks, DOE officials may not have the necessary information needed to inform NASA about the effects of these challenges. These potential effects may result in DOE not meeting its Pu-238 production goal, which could cause NASA to delay future RPS-powered missions. DOE officials said that they do not maintain a risk tracking system for the RPS program and that this is maintained by NASA. However, according to NASA officials we interviewed, NASA only tracks high-level risks that could directly affect NASA. *Standards for Internal Control in the Federal Government* states that agency management should identify, analyze, and respond to risks related to achieving defined objectives. Management is to identify risk throughout the agency to provide a basis for analyzing risks. With a more comprehensive risk tracking system that allows DOE to identify more systemic risks beyond the specific technical risks identified by individual laboratories, DOE will be better positioned to adequately assess the longterm effects of all of the identified challenges associated with Pu-238 and RPS production objectives.

CONCLUSION

NASA is entirely dependent on DOE to supply Pu-238 and RPS for space missions, and future missions requiring RPS are at risk if DOE is unable to fuel and provide these RPS. To meet NASA's demand for new Pu-238, DOE's Supply Project goal of 1.5 kg per year was initially established assuming the use of a more efficient RPS design, which has yet to be realized. While NASA officials subsequently said that the full production rate under DOE's Supply Project would be sufficient using existing RPS designs, any Pu-238 production shortage, coupled with the use of less efficient RPS technology, could present a challenge to carrying out NASA's future RPS-powered missions. In addition, if any of NASA's Pu238 supply is needed for national security or other applications, NASA may not have sufficient Pu-238 to support future missions or will have to delay such missions until more Pu-238 is provided under the Supply Project. Finally, if NASA intends to launch an additional RPS-powered mission before the end of the 2020s, NASA officials would have to make such a decision by 2023, before full-scale Pu-238 production under the Supply Project is achieved, because it takes multiple years between the time a mission and its power source are selected to the time of launch.

DOE has taken steps to reestablish domestic production of Pu-238 under its Supply Project, and has produced small quantities of new Pu-238. In addition, DOE has identified key challenges to the Supply Project—such as scaling up chemical processing and qualifying targets—that put achieving Supply Project production goals at risk. However, DOE has not fully assessed the potential long-term effects of these challenges on production goals. While DOE officials anticipate that their new approach to managing RPS and Pu-238 production will better help address these challenges, DOE is in the very early stages of implementing this approach and has not identified details, including milestones and interim steps, for how it would address them. In addition, DOE's plans for nuclear weapons pit production activities have not taken into account their potential effects on RPS production and the Supply Project. Federal standards for internal control state that agency management should use and communicate quality information externally to help the entity achieve its objectives and address related risks. While DOE and its laboratories have identified existing challenges to RPS and Pu-238 production, the effect of these challenges on expected Pu-238 production goals have not been fully assessed and thus not fully communicated with NASA. Moreover, DOE's risk tracking system for RPS does not track systemic risks, but rather relies on individual laboratories

INL, the laboratory primarily responsible for overseeing RPS production, to meet with officials and examine facilities involved in the Supply Project and RPS production. Finally, we compared DOE's efforts to collect and assess quality information about challenges associated with RPS and Pu238 production and to communicate these challenges against criteria outlined in federal internal control standards.[5] We also evaluated DOE's management approach for RPS and Pu-238 production against key management practices established in prior GAO work.[6] We conducted this performance audit from March 2016 to September 2017 in accordance with generally accepted government auditing standards. Those standards require that we plan and perform the audit to obtain sufficient, appropriate evidence to provide a reasonable basis for our findings and conclusions based on our audit objectives. We believe that the evidence obtained provides a reasonable basis for our findings and conclusions based on our audit objectives.

APPENDIX II: NATIONAL ACADEMY OF SCIENCES DECADAL SURVEY RECOMMENDED MISSIONS AND POWER SOURCES, 2013-2022

Table 3. National Academy of Sciences' (NAS) 2013-2022 Decadal Survey Recommended Missions and Power Sources

Mission	Mission Type (mission number)	Power Source	Power Source Type (number of units proposed)
Mars Astrobiology Explorer-Cacher (MAX-C)[a]	Flagship	Solar	Ultraflex (2)
Jupiter Europa Orbiter[b]	Flagship	RPS	MMRTG (5)
Uranus Orbiter and Probe	Flagship	RPS	ASRG (1)
Comet Surface Sample Return	New Frontiers (4 or 5)	Solar	Ultraflex (1)

[5] GAO, Standards for Internal Control in the Federal Government, GAO-14-704G (Washington, D.C.: September 2014).

[6] GAO, Defense Health Care Reform: Actions Needed to Help Ensure Defense Health Agency Maintains Implementation Progress, GAO-15-759 (Washington, D.C.: Sept. 10, 2015), and Biobased Products: Improved USDA Management Would Help Agencies Comply with Farm Bill Purchasing Requirements, GAO-04-437 (Washington, D.C.: Apr. 7, 2004).

Mission	Mission Type (mission number)	Power Source	Power Source Type (number of units proposed)
Lunar South Pole-Aitken Basin Sample Return[c]	New Frontiers (4 or 5)	RPS	Not specified
Saturn Probe	New Frontiers (4 or 5)	RPS	ASRG (2)
Trojan Tour and Rendezvous	New Frontiers (4 or 5)	RPS	ASRG (2)
Venus In Situ Explorer[d]	New Frontiers (4 or 5)	Solar	Advanced solar-electric propulsion
Io Observer	New Frontiers (5)	RPS	ASRG (2)
Lunar Geophysical Network	New Frontiers (5)	RPS	ASRG (1)
Venus Climate Mission	Beyond the current decade	Solar	Gimbaled (1)
Enceladus Orbiter	Beyond the current decade	RPS	ASRG (3)
Mars Sample Return Lander and Mars Ascent Vehicle	Beyond the current decade	Solar	Ultraflex (1) + Fetch Rover (1)
Mars Sample Return Orbiter and Earth Entry Vehicle	Beyond the current decade	Solar	Ultraflex (1)
Titan Saturn System Mission	Beyond the current decade	RPS	ASRG (5) + MMRTG (1)
Neptune System Orbiter and Probe	Beyond the current decade	RPS	ASRG (3 to 6)

Legend: RPS = radioisotope power system; MMRTG = Multi-Mission Radioisotope Thermoelectric Generator; ASRG = Advanced Stirling Radioisotope Generator.

Source: NAS Decadal Survey | GAO-17-673.

Notes: The NAS decadal survey report, Vision and Voyages for Planetary Science in the Decade 2013-2022, proposes an array of missions to be considered by the National Aeronautics and Space Administration (NASA). The decadal survey report acknowledges that NASA cannot carry out all of these missions. The decadal survey lists Flagship missions in priority order. The New Frontiers missions are not prioritized, and Discovery missions are not listed in the survey report. The survey report also suggests technological investments for some missions beyond the current decade.

[a] NASA has chosen to pursue the Mars 2020 Flagship mission in this decade as an alternative to the MAX-C mission. Mars 2020 is in the planning and development phase and will use one RPS, rather than solar power, which is proposed for the MAX-C mission in the decadal survey report.

[b] NASA is pursuing the Europa Clipper mission, originally called the Jupiter Europa Orbiter (JEO) mission, which is in the planning and development phase. Although originally proposed as an RPS powered mission, NASA determined that solar power is a feasible power source for the Europa Clipper.

[c] The Lunar South Pole-Aitken Basin Sample Return was originally proposed in the 2003-2013 decadal survey report.

[d] The Venus In Situ Explorer mission was originally proposed in the 2003-2013 decadal survey report.

APPENDIX III: COMMENTS FROM THE DEPARTMENT OF ENERGY

Department of Energy
Washington, DC 20585

August 14, 2017

Ms. Shelby S. Oakley
Director, Natural Resources
 and Environment
U.S. Government Accountability Office
441 G Street, NW
Washington, D.C. 20548

Dear Ms. Oakley:

Thank you for providing a draft copy of the Government Accountability Office (GAO) Report "Space Exploration: DOE Could Improve Planning and Communication Related to Plutonium-238 and Radioisotope Power Systems Production Challenges" (GAO-17-673). We appreciate GAO's efforts in this review.

The Office of Nuclear Energy (NE) is in the process of implementing a new approach for the Radioisotope Power Systems (RPS) supply chain that is more responsive to the National Aeronautics and Space Administration's (NASA) needs. This new strategy, termed Constant Rate Production (CRP), will position the RPS infrastructure to support a sustained level of production in line with production goals of NASA. The CRP strategy includes an approach to identifying and assessing issues that could have long term effects to program success as well as communication of issues to NASA. This strategy will be executed through an integrated program plan that prioritizes activities across the program laboratories while linking to risks identified by both Agencies.

The draft report recommendations are:

Recommendation 1: Develop an implementation plan with milestones and interim steps for the department's management approach for Pu-238 and RPS production.

We concur with the recommendation. The Department is developing an integrated program plan to implement the Constant Rate Production strategy that will document the management approach for Pu-238 and RPS production. Estimated completion date: September 2018.

Recommendation 2: Assess the long-term effects that known challenges may have on production quantities, time frames, or required funding, and communicate these potential effects to NASA.

We concur with the recommendation. The Department will work with NASA to identify, assess, and develop plans to address known challenges to the RPS program. Estimated completion date: September 2019.

Recommendation 3: Develop a more comprehensive system to track more systemic risks, beyond the specific technical risks identified by individual laboratories.

We concur with the recommendation. As part of the integrated program plan, DOE will include steps to ensure the current risk system includes comprehensive programmatic risks to support NASA's risk management activities. Estimated completion date: September 2018.

If you have any questions, please contact Kelly Scott at kelly.scott@nuclear.energy.gov or 202-586-4288.

Sincerely,

Ed McGinnis,
Acting Assistant Secretary
for Nuclear Energy

APPENDIX IV: STAFF ACKNOWLEDGMENTS

In addition to Shelby S. Oakley, Jonathan Gill (Assistant Director), Samuel Blake, Kevin Bray, John Delicath, Jennifer Echard, Cindy Gilbert, Timothy Guinane, John Hocker, Michael Kaeser, Jason Lee, Danny Royer, Aaron Shiffrin, Kiki Theodoropoulos, Kristin VanWychen, and John Warren made key contributions to the work.

APPENDIX V: ACCESSIBLE DATA

Data Tables

Data for Figure 1: Expanded View of the Multi-Mission Radioisotope Thermoelectric Generator (MMRTG)

Illustration of Multi-Mission Radioisotope Thermoelectric Generator (expanded view) identifying these components:
<ul><li>General-purpose heat source (GPHS) module</li><li>Stack of eight GPHS modules</li><li>Thermoelectric module (thermocouples)</li><li>Radiator fins</li></ul>

Data for Figure 2: Overview of the Plutonium-238 (Pu-238) Supply Project and Radioisotope Power System (RPS) Production Process

---

Plutonium-238 (Pu-238) Supply Project

    1)   Idaho National Laboratory Neptunium-237 (Np-237) storage
       a.   Receives Recycled Np-237 from step 5, Oak Ridge National Laboratory
    2)   Oak Ridge National Laboratory Np-237 pellet pressing and placement into targets
    3)   Proceeds to either:
       a.   Oak Ridge National Laboratory
       b.   Np-237 target irradiation at High Flux Isotope Reactor Conversion of Np-237 to Pu-238 isotope
       c.   Idaho National Laboratory Np-237 target irradiation at Advanced Test Reactor Conversion of Np-237 to Pu-238 isotope planned
    4)   Oak Ridge National Laboratory Chemical processing Np-237 and Pu-238 isotope separation Creation of Pu-238 oxide

Radioisotope power system (RPS) production

    1.   Idaho National Laboratory
       •   RPS assembly and testing
       •   RPS storage and delivery
    2.   2. Idaho National Laboratory
       •   GPHS module fabrication
       •   GPHS module storage
    3.   Los Alamos National Laboratory
       •   Pu-238 general-purpose heat source (GPHS) fuel clad production
    4.   4. Los Alamos National Laboratory
    Pu-238 oxide storage

Source: GAO analysis of Department of Energy documents. | GAO-17-673.

Data for Figure 4: Power Source Selection in NASA's Lifecycle Review Process

---

Pre-phase A Concept studies, Power source selection made and approved

    •   Phase A, Concept and technology development
    •   Phase B, Preliminary design and technology competition

Data for Figure 4 (Continued)

•   Phase C, Final design and fabrication
•   Phase D, System assembly, integration and test, launch
•   Phase E, Operations and sustainment
•   Phase F, Closeout

Source: GAO analysis of National Aeronautics and Space Administration guidance. | GAO-17-673.

Timeline for Figure 5: Selected Chronology of Department of Energy and National Aeronautics and Space Administration Radioisotope Power Systems and Plutonium-238 Production Activities

DOE's RPS and Pu-238 key milestones
2011: Supply Project initiated
2016: First Pu-238 production run
2019: Interim production rate (300-500 g/year)
2023: Early finish of 1.5 kg Pu-238 production rate
2026: Late finish of 1.5 kg Pu-238 production rate
NASA mission-related activities
2015: Mars 2020 RPS production
2019: New Frontiers #4 mission selection
2020: Mars 202 launch
2025: New Frontiers #4 launch
2029: Earliest launch of one MMRTG using new Pu-238
2030: Earliest launch of two MMRTG using new Pu-238
Expected milestones and activities
Existing Pu-238 supply exhausted with launch of New Frontiers #4 if three RPS are used
Abbreviations:
DOE, Department of Energy
MMRTG, Multi-Mission Radioisotope Thermoelectric Generator
NASA, National Aeronautics and Space Administration
Pu-238, plutonium-238
RPS, radioisotope power system

Source: GAO analysis of DOE and NASA information. | GAO-17-673.

AGENCY COMMENT LETTER

Text of Appendix III: Comments from the Department of Energy

August 14, 2017

Ms. Shelby S. Oakley Director, Natural Resources and Environment U.S. Government Accountability Office 441 G Street, NW

Washington, D.C. 20548

Dear Ms. Oakley:

Thank you for providing a draft copy of the Government Accountability Office (GAO) Report "Space Exploration: DOE Could Improve Planning and Communication Related to Plutonium-238 and Radioisotope Power Systems Production Challenges" (GAO-17- 673). We appreciate GAO's efforts in this review.

The Office of Nuclear Energy (NE) is in the process of implementing a new approach for the Radioisotope Power Systems (RPS) supply chain that is more responsive to the National Aeronautics and Space Administration's (NASA) needs. This new strategy, termed Constant Rate Production (CRP), will position the RPS infrastructure to support a sustained level of production in line with production goals of NASA. The CRP strategy includes an approach to identifying and assessing issues that could have long term effects to program success as well as communication of issues to NASA. This strategy will be executed through an integrated program plan that prioritizes activities across the program laboratories while linking to risks identified by both Agencies.

The draft report recommendations are:

Recommendation 1: Develop an implementation plan with milestones and interim steps for the department's management approach for Pu-238 and RPS production.

We concur with the recommendation. The Department is developing an integrated program plan to implement the Constant Rate Production strategy that will document the management approach for Pu-238 and RPS production. Estimated completion date: September 2018.

Recommendation 2: Assess the long-term effects that known challenges may have on production quantities, time frames, or required funding, and communicate these potential effects to NASA.

We concur with the recommendation. The Department will work with NASA to identify, assess, and develop plans to address known challenges to the RPS program. Estimated completion date: September 2019.

Recommendation 3: Develop a more comprehensive system to track more systemic risks, beyond the specific technical risks identified by individual laboratories.

We concur with the recommendation. As part of the integrated program plan, DOE will include steps to ensure the current risk system includes comprehensive programmatic risks to support NASA's risk management activities. Estimated completion date: September 2018.

Ed McGinnis,
Acting Assistant Secretary for Nuclear Energy

In: Space Exploration
Editor: Thomas R. Reed

Chapter 4

SPACE EXPLORATION: IMPROVED PLANNING AND COMMUNICATION NEEDED FOR PLUTONIUM-238 AND RADIOISOTOPE POWER SYSTEMS PRODUCTION[*]

Shelby S. Oakley

Chairman Babin, Ranking Member Bera, and Members of the Subcommittee:

I am pleased to be here today to discuss our recent work on radioisotope power systems. The National Aeronautics and Space Administration (NASA) has long used radioisotope power systems (RPS) to generate reliable electrical power and heat energy for long-duration space missions. RPS can

---

[*] This is an edited, reformatted and augmented accessible version of the United States Government Accountability Office Testimony Before the Subcommittee on Space, Committee on Science, Space, and Technology, House of Representatives, Publication No. GAO-18-161T, dated October 4, 2017.

operate where solar panels or batteries would be ineffective or impossible to use, such as in deep space or in shadowed craters, by converting heat from the natural radioactive decay of plutonium-238 (Pu-238) into electricity.[1] The Department of Energy (DOE) and its predecessor agencies have been providing Pu-238 and fabricating RPS for NASA and other federal agencies for more than 5 decades.[2]

In 2011, with funding provided by NASA, DOE initiated the Pu-238 Supply Project (Supply Project) to reestablish the capability to domestically produce Pu-238.[3] According to DOE documents and agency officials, DOE currently maintains about 35 kilograms (kg) of Pu-238 isotope designated for NASA missions, about half of which currently meets the power specifications for spaceflight. However, given NASA's current plans for solar system exploration, this supply could be exhausted within the next decade. Specifically, NASA plans to use about 3.5 kg of Pu-238 isotope for one RPS to power the Mars 2020 mission. NASA may also use an additional 10.5 kg of Pu-238 isotope for its New Frontiers #4 mission if three RPS are used.[4] If DOE's existing Pu-238 supply is used for these two missions, NASA would be forced to eliminate or delay future missions requiring RPS until DOE produces or acquires more Pu-238.

My remarks today are based on our recent report on NASA's use of radioisotope power systems that are powered by plutonium 238,[5] which we are releasing today. Our report examined (1) how NASA selects RPS for missions and what factors affect NASA's demand for RPS and Pu238; and

---

[1] Pu-238 is defined as Pu-238 oxide, also known as "heat source" plutonium oxide or "bulk-oxide", and is the form used to power RPS. Pu-238 isotope is a precursor to Pu-238 oxide.

[2] The Atomic Energy Act of 1954 authorizes DOE to provide systems that meet the special nuclear material needs of other federal agencies. Under a 1991 agreement with NASA, which was revised in 2016, DOE is responsible for maintaining our nation's capability to support the development, production, and safety of NASA's space exploration missions that use RPS.

[3] Historically, Pu-238 was produced domestically or was purchased from Russia. Domestic Pu-238 production ended in 1988, and DOE has not purchased material from Russia since 2009.

[4] NASA has offered up to 3 RPS for the New Frontiers Mission and plans to make a mission selection in July 2019.

[5] GAO, Space Exploration: DOE Could Improve Planning and Communication Related to Plutonium-238 and Radioisotope Power Systems Production Challenges, GAO-17-673 (Washington, D.C.; Sept. 8, 2017).

(2) DOE's progress in meeting NASA's RPS and Pu-238 demand, and what, if any, challenges DOE faces in meeting the demand. Today, I will discuss the key findings and recommendations from that report.

For our report, we reviewed documentation on how NASA considered mission requirements during the agency's planning for recent missions that considered or used RPS as a power source. We also interviewed officials from the Planetary Science Division (PSD) of NASA's Science Mission Directorate and from the Human Exploration and Operations Mission Directorate. In addition, we reviewed documentation related to DOE's efforts to develop the Supply Project and DOE's RPS production process. We also interviewed officials from DOE's Office of Nuclear Energy as well as officials involved in RPS work at three DOE national laboratories—Idaho National Laboratory (INL), Los Alamos National Laboratory (LANL), and Oak Ridge National Laboratory (ORNL)—and conducted site visits to ORNL and INL. More detailed information on the objectives, scope, and methodology of our work can be found in the September report. The work upon which this statement is based was conducted in accordance with generally accepted government auditing standards.

In summary, we found that NASA selects RPS for missions based primarily on scientific objectives and that several factors may affect NASA's demand for RPS and Pu-238. For example, RPS have been typically used on NASA's most expensive and highest priority missions. Based on expected funding levels, NASA can only support two or three of these missions per decade. We also found that DOE has made progress meeting NASA's demand for RPS and Pu-238, but the agency faces some challenges in reaching full production goals. For example, DOE does not maintain a comprehensive system for tracking RPS production risks. In addition, DOE's management approach does not allow for the agency to adequately communicate long-term production challenges to NASA. We made several recommendations to DOE to address these issues. DOE agreed with our recommendations and outlined actions it planned to take to address them.

BACKGROUND

RPS are long-lived sources of spacecraft electrical power and heating that are rugged, compact, highly reliable, and relatively insensitive to radiation and other effects of the space environment, according to NASA documentation. Such systems can provide spacecraft power for more than a decade and can do so billions of miles from the sun. Twenty-seven U.S. missions have used RPS over the past 5 decades. The current RPS design, the Multi-Mission Radioisotope Thermoelectric Generator (MMRTG), converts heat given off by Pu-238 into about 120 watts of electrical power at the beginning of its life—a 6 percent power conversion efficiency. One MMRTG contains 32 general purpose heat source (GPHS) fuel clads in the form of pressed Pu-238 pellets encapsulated in iridium.

NASA's PSD science portfolio includes a wide array of missions that seek to address a variety of scientific objectives and answer many questions about the solar system, from how life began to how the solar system is evolving. Scientific and mission objectives influence the types of equipment needed for the mission, including the mission's power source. According to NASA officials we interviewed, missions in NASA's PSD portfolio are generally classified in three ways:

- *Flagship.* Flagship missions are the largest and most expensive class of NASA's missions, costing $2 billion or more, and are given the highest priority for resources, including funding, infrastructure, and launch support. Past Flagship missions that have used RPS include the Galileo, Cassini, and Curiosity missions. NASA's Mars 2020 mission is a planned Flagship mission using RPS.
- *New Frontiers.* New Frontiers missions focus on enhancing our understanding of the solar system and have a development cost cap of $850 million.[6] To date, there has been one New Frontiers mission using RPS (New Horizons).

---

[6] Mission cost caps are in fixed fiscal year 2015 dollars and do not include certain costs, such as those related to the launch vehicle and operations.

- *Discovery.* Missions in the Discovery program have a development cost cap of $450 million to $500 million and have shorter development time frames, according to NASA officials and documentation. No Discovery mission has been powered by RPS.

DOE oversees the design, development, fabrication, testing, and delivery of RPS to meet NASA's overall systems requirements, specifications, and schedules. DOE's goal under its Supply Project is to reach a full Pu-238 production rate of 1.5 kg per year by 2023, at the earliest, with a late completion date of 2026. DOE also established an interim production rate of 300 to 500 grams per year by 2019, to ensure an adequate supply of Pu-238 for NASA's near-term missions, before the full production rate goal is achieved. The Supply Project involves a number of steps across several DOE national laboratories, including the use of two DOE research reactors—the High Flux Isotope Reactor at ORNL, and the Advanced Test Reactor at INL.

NASA began fully funding DOE's Supply Project in 2011, and since 2014, has been responsible for funding all aspects of RPS production operations, according to NASA documents.[7] NASA funds DOE's efforts to build, test, and fuel RPS, as well as to update equipment and sustain staffing levels associated with RPS production between missions. Since 2014 NASA has provided, on average, approximately $50 million per year to support DOE's ongoing operations and maintenance of RPS production equipment. Since its inception until early 2017, DOE has used a short-term and incremental segmented management approach to manage the Supply Project.[8]

---

[7] Prior to 2014, DOE provided funding for the infrastructure related to RPS production at DOE facilities, and NASA provided funding for mission-specific RPS production activities.

[8] Under this management approach, DOE established short-term segments of Supply Project work to be connected to time frames over which DOE could more reliably predict funding from NASA.

NASA Selects RPS for Missions Based Primarily on Scientific Objectives, and Several Factors May Affect NASA's Demand for RPS and Pu-238

NASA selects RPS to power its missions primarily based on scientific objectives and mission destinations. According to NASA officials we interviewed, the need for RPS is usually apparent based on the mission's scientific objectives and destination. For instance, an RPS is more likely to be needed for a mission to a distant planet, where minimal sunlight reduces the effectiveness of solar power. NASA officials we interviewed stated that, consistent with the National Space Policy, the agency uses RPS when they enable or significantly enhance a mission, or when alternative power sources, such as solar power, might significantly compromise mission objectives.[9] NASA prioritizes mission selection based on missions identified in the National Academy of Sciences' decadal survey report, which represents the highest priorities of the scientific community and includes many missions that require the use of RPS.[10]

Prior to the establishment of DOE's Supply Project in fiscal year 2011, NASA officials we interviewed stated that mission selections were influenced by the limited amount of available Pu-238. These same officials told us that missions are now selected independently from decisions about how they will be powered. However, projected availability of Pu-238 is factored into whether an RPS is available for a specific mission opportunity.

In addition to the scientific objectives of planned and potential space exploration missions, several other factors may affect NASA's demand for RPS and Pu-238:

---

[9] U.S. Office of Science and Technology Policy, National Space Policy of the United States of America (Washington, D.C.: June 28, 2010).

[10] The National Academy of Sciences' decadal survey report presents a 10-year program of science and exploration with the potential to yield revolutionary new discoveries. The National Aeronautics and Space Administration Transition Authorization Act of 2017 states that the NASA Administrator should set science priorities by following guidance provided in this decadal survey report.

- *Costs associated with missions that typically require RPS.* According to NASA officials, RPS have typically been used on Flagship missions that cost $2 billion or more. NASA can support no more than one mission using RPS about every 4 years—or two to three missions per decade—based on expected agency funding levels.
- *Cost of RPS relative to mission costs.* According to NASA officials, New Frontiers missions may be good candidates to use RPS; however, given the cost cap for this mission class, one RPS would account for about 9 percent of the mission's budget, while three RPS would account for almost 14 percent. For Discovery missions, for which the cost of using RPS would represent a large portion of a Discovery mission budget, a single RPS would represent more than 17 percent of a mission's development cap.
- *DOE's production capability.* According to DOE officials we interviewed, it can take up to 6 years to acquire, fuel, test, and deliver a new RPS for a NASA mission. According to DOE and NASA officials we interviewed, given the current floor space dedicated to RPS development at INL and limits on staff exposure to radiation at LANL, DOE only has the capacity to produce three to four RPS at one time. To accommodate DOE's current RPS production capability, NASA officials we interviewed said they will not select two consecutive missions requiring RPS.

Technological advances may reduce the demand for Pu-238 and thus RPS. For example, according to NASA officials, advances in solar power technology have realistically expanded the ability to use solar power for missions for which it would not have been considered before, and these advances could help address low levels of light intensity for deep space missions. NASA also is developing new RPS technologies that may reduce its demand for Pu-238 and thus RPS. For example, NASA officials told us that they plan to invest in dynamic RPS technology that could increase RPS efficiency and require less RPS to achieve mission power. NASA research indicates that dynamic RPS designs could be more than four times as efficient as the current MMRTG design.[11]

The Supply Project goal of producing 1.5 kg of Pu-238 per year was established to support two to three PSD missions using RPS each decade, and NASA does not anticipate other potential users to affect demand for RPS or Pu-238, according to NASA and DOE officials and documentation we reviewed. DOE planning documents and NASA officials we interviewed stated that current RPS and Pu-238 production levels expected from the Supply Project are intended to only meet PSD's demand. NASA officials said that they did not account for potential demand from other potential users within NASA, the national security community, or commercial sectors when establishing current Pu-238 production goals.

DOE HAS MADE PROGRESS MEETING NASA'S RPS AND PU-238 DEMAND, BUT FACES CHALLENGES REACHING FULL PRODUCTION GOALS

DOE has made progress meeting NASA's future demand for Pu-238 to fuel RPS. A chronology of key DOE planned RPS and Pu-238 production activities, and NASA's mission-related activities, are shown in figure 1.

DOE demonstrated a proof of concept for new Pu-238 production, and has made approximately 100 grams of new Pu-238 isotope under its Supply Project, since the project's inception in 2011. However, given DOE's Supply Project and RPS production schedule, and NASA's current space exploration plans to use up to four RPS for its Mars 2020 and New Frontiers #4 missions, DOE's existing Pu-238 supply will be exhausted by 2025.

---

[11] In addition, new thermoelectric materials being researched by NASA may lead to increased RPS efficiency. One such material, skutterudite, could result in an RPS with a 25 percent beginning-of-life efficiency improvement and a 50 percent increase of end-oflife power output when compared to the current MMRTG.

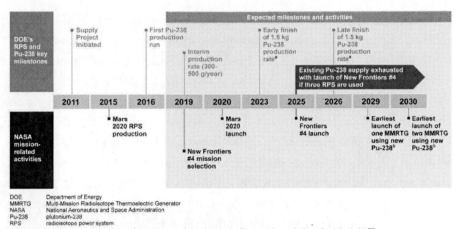

DOE   Department of Energy
MMRTG   Multi-Mission Radioisotope Thermoelectric Generator
NASA   National Aeronautics and Space Administration
Pu-238   plutonium-238
RPS   radioisotope power system

Source: GAO analysis of DOE and NASA information. | GAO-18-161T.

[a] DOE has established a date range of 2023 to 2026 to achieve full production of 1.5 kg of Pu-238 per year; however, as of June 2017, DOE officials said they expect to reach full production no earlier than 2025.

[b] Earliest launch dates assume delivery of 1.5 kg of Pu-238 by the start of fiscal year 2026.

Figure 1. Chronology of Key Department of Energy and National Aeronautics and Space Administration Radioisotope Power Systems and Plutonium-238 Production Activities.

Moreover, DOE officials we interviewed from INL, LANL, and ORNL identified several challenges, including perfecting and scaling up chemical processing and the availability of reactors, that need to be overcome for DOE to meet its projected Supply Project goal of producing 1.5 kg per year of Pu-238 by 2026, at the latest. If these challenges are not overcome, DOE could experience delays in producing Pu-238 to fuel RPS for future NASA missions.

DOE's ability to meet its production goal and support future NASA missions is at risk if certain steps for chemical processing necessary for the production of Pu-238 are not improved and scaled up. According to DOE officials we interviewed, DOE is still in the experimental stage and has not perfected certain chemical processing measures required to extract new Pu-238 isotope from irradiated targets, creating a bottleneck in the Supply Project and putting production goals at risk.

In addition, reactor availability will be necessary for DOE to achieve its Pu-238 production goals. Officials we interviewed at INL and ORNL said that achieving 1.5 kg of Pu-238 per year is contingent on the availability of positions within both the High Flux Isotope Reactor (HFIR) and the Advanced Test Reactor (ATR) to irradiate neptunium targets for conversion to Pu-238 isotope.[12] DOE officials said HFIR can produce approximately 600 grams of Pu-238 isotope and they plan to use positions within ATR to achieve full production goals; however, ATR has not been qualified for Supply Project work. In addition, DOE officials said that ATR's availability for the Supply Project may be limited due to competition from other users. DOE officials said that they will be unable to meet full Pu-238 production goals if positions in ATR, which are already over-utilized, are not available for Pu-238 isotope production and that they do not have a plan to address this challenge.

These and other challenges identified in our September 2017 report may place DOE's RPS and Pu-238 production goals at risk, in part, because of the short-term and incremental segmented management approach DOE had used to manage the Supply Project since its inception in 2011 through early 2017. In March 2017, DOE officials we interviewed said that the agency anticipated moving to a constant GPHS production rate approach to help provide funding flexibility and stabilize RPS production staffing levels between NASA missions. In June 2017, DOE officials we interviewed said that implementing a constant GPHS production rate approach would also address other previously identified challenges associated with RPS production and the Supply Project and therefore decided to discontinue its short-term and incremental segmented management approach.

However, DOE officials we interviewed did not describe how the new constant GPHS production rate approach would help them address some of the longer-term challenges previously identified by the agency, such as scaling up and perfecting chemical processing. We found that DOE has yet

---

[12] Positions are locations within the reactors where targets are bundled and placed for the irradiation process. Only certain positions are suitable for Pu-238 production. According to DOE documentation, HFIR has 22 positions within the reactor, 20 of which are suitable for Pu-238 isotope production. According to INL documentation and officials we interviewed, ATR has 75 positions within the reactor, of which 9 are suitable for Pu-238 isotope production.

to develop an implementation plan for the new approach, with defined tasks and milestones, that can be used to show progress toward assessing challenges, demonstrate how risks are being addressed, or assist in making adjustments to its efforts when necessary. Our previous work has shown that without defined tasks and milestones, it is difficult for agencies to set priorities, use resources efficiently, and monitor progress toward achieving program objectives.[13]

In our September 2017 report, we recommended that DOE develop a plan that outlined interim steps and milestones that would allow the agency to monitor and assess the implementation of its new approach for managing Pu-238 and RPS production. DOE agreed with our recommendation and noted it was in the process of implementing an approach for the RPS supply chain that was more responsive to NASA's needs, among other things. DOE also noted that it was developing an integrated program plan to implement and document the agency's new approach and expected this to be completed in September 2018. We believe that the development of an integrated program plan is an important step and that any such plan should include defined tasks and milestones, so that DOE can demonstrate progress toward achieving its RPS supply chain goals.

In addition, in our September 2017 report we identified another factor that could undermine DOE's ability to inform NASA about previously identified challenges to reach its full Pu-238 production goal. We found that DOE does not maintain a comprehensive system for tracking RPS production risks and, instead, relies on individual laboratories to track and manage risks specific to their laboratories. *Standards for Internal Control in the Federal Government* call for agency management to identify, analyze, and respond to risks related to achieving defined objectives.[14] We recommended that DOE develop a more comprehensive system to track systemic risks, beyond the specific technical risks identified by individual laboratories. Doing so would

---

[13] GAO, Defense Health Care Reform: Actions Needed to Help Ensure Defense Health Agency Maintains Implementation Progress, GAO-15-759 (Washington, D.C.: Sept. 10, 2015), and Biobased Products: Improved USDA Management Would Help Agencies Comply with Farm Bill Purchasing Requirements, GAO-04-437 (Washington, D.C.: Apr. 7, 2004).

[14] GAO, Standards for Internal Control in the Federal Government, GAO-14-704G (Washington, D.C.: September 2014).

better position DOE to assess the long-term effects of the challenges associated with its Pu-238 and RPS production objectives. DOE agreed with our recommendation and stated that the agency would include steps to ensure that its risk assessment system would include comprehensive programmatic risks.

Finally, in our September 2017 report we found that DOE's new approach to managing RPS and Pu-238 production does not allow for DOE to adequately communicate long-term challenges to NASA. It is also unclear how DOE will use its new management approach to communicate to NASA challenges related to Pu-238 production. As a result, NASA may not have adequate information to plan for future missions using RPS. *Standards for Internal Control in the Federal Government* call for agency management to use quality information to achieve agency objectives and communicate quality information externally through reporting lines so that external parties can help achieve agency objectives and address related risks. In our September 2017 report, we recommended that DOE assess the long-term effects that known challenges may have on Pu-238 production quantities, time frames, and required funding, and communicate these potential effects to NASA. DOE stated that it agreed with our recommendation and would work with NASA to identify, assess, and develop plans to address known challenges. DOE also stated that the agency expected to complete this effort in September 2019.

Chairman Babin, Ranking Member Bera, and Members of the Subcommittee, this concludes my prepared statement. I would be pleased to respond to any questions that you may have at this time.

STAFF ACKNOWLEDGMENTS

Individuals who made key contributions to the report on which this testimony is based are Shelby Oakley, Jonathan Gill (Assistant Director); Samuel Blake, Kevin Bray, John Delicath, Jennifer Echard, Cindy Gilbert,

Timothy Guinane, John Hocker, Michael Kaeser, Jason Lee, Tim Persons, Danny Royer, Aaron Shiffrin, Kiki Theodoropoulos, Kristin VanWychen, and John Warren.

INDEX

to track and manage specific risks. Better information about risks would allow DOE officials to inform NASA about the effects of these challenges. Supply Project challenges, if not addressed, may result in DOE not meeting, or being delayed in meeting, the Supply Project goals, and DOE may not be able to fully support NASA missions to deep space using RPS after New Frontiers #4 in 2025.

RECOMMENDATIONS FOR EXECUTIVE ACTION

To help ensure the availability of Pu-238 and RPS for space exploration, we recommend that the Secretary of Energy take the following three actions:

- develop an implementation plan with milestones and interim steps for the department's management approach for Pu-238 and RPS production;
- assess the long-term effects that known challenges may have on production quantities, time frames, or required funding, and communicate these potential effects to NASA; and
- develop a more comprehensive system to track more systemic risks, beyond the specific technical risks identified by individual laboratories.

AGENCY COMMENTS AND OUR EVALUATION

We provided a draft of this chapter to DOE and NASA for review and comment. In response, we received written comments from DOE, which are summarized below and reprinted in appendix III. NASA did not provide a formal response because the report made no recommendations to NASA; instead, NASA provided technical comments, which we incorporated as appropriate. DOE concurred with our three recommendations, stating that it is in the process of implementing a new approach, termed Constant Rate Production, for the RPS supply chain that is more responsive to NASA's needs, among other things. In response to our first recommendation, DOE said that the agency is developing an integrated program plan to implement the Constant Rate Production strategy that will document the management approach for Pu-238 and RPS production. DOE expects to complete this plan in September 2018. We believe that the development of an integrated program plan is an important step, and reiterate that any such plan documenting DOE's management approach should include milestones and interim steps so DOE can show progress toward implementation, among other things. In response to our second recommendation, DOE said that the agency will work with NASA to identify, assess, and develop plans to address known challenges to the RPS program and expects to complete this effort in September 2019. In response to our third recommendation, DOE said that as part of the integrated program plan it intends to develop in response to our first recommendation, the agency will include steps to ensure the current risk system includes comprehensive programmatic risks to support NASA's risk management activities. We believe that implementing these actions will help DOE better ensure the availability of Pu-238 and RPS in support of NASA's space exploration missions.

As agreed with your offices, unless you publicly announce the contents of this chapter earlier, we plan no further distribution until 30 days from the report date. At that time, we will send copies to the appropriate congressional committees, the Secretary of Energy, the Administrator of the National Aeronautics and Space Administration, and other interested parties.

Shelby S. Oakley

Shelby S. Oakley Director
Acquisition and Sourcing Management

APPENDIX I: OBJECTIVES, SCOPE, AND METHODOLOGY

This chapter examines the National Aeronautics and Space Administration's (NASA) process for considering and selecting power sources for missions, in particular the use of radioisotope power systems

(RPS), and the Department of Energy's (DOE) ability to maintain the necessary infrastructure and workforce for RPS and plutonium-238 (Pu238) production. Our objectives were to (1) describe how NASA selects RPS for missions and what factors affect RPS and Pu-238 demand; and (2) evaluate DOE's progress in meeting NASA's RPS and Pu-238 demand and what challenges, if any, DOE faces in meeting the demand.

To describe how NASA selects RPS for missions and what factors affect RPS and Pu-238 demand, we reviewed NASA documents, including procedural requirements, key decision documentation for specific missions, and agreements between NASA and DOE. Specifically, we reviewed NASA's procedural requirements, including the *NASA Space Flight Program and Project Management Handbook*, and reviewed how those requirements were applied to recent missions that used or considered RPS as a power source.[1] We also reviewed documentation related to NASA's key decisions for specific missions—including Mars Space Laboratory, Mars 2020, and the Europa Clipper—to identify NASA's power source decisions and when they were made under NASA's review process. To determine how NASA and DOE collaborate on RPS and Pu-238 development, we reviewed interagency agreements and memoranda of understanding between these two agencies. We also met with NASA officials to discuss how mission decisions related to power sources are made and the impact of technological advances on NASA's demand for RPS. In addition, we conducted interviews with officials from the Planetary Science Division (PSD) of NASA's Science Mission Directorate to determine how NASA selects its missions and how, if at all, the availability of Pu-238 has affected mission selection. In addition, we asked PSD officials about the process NASA uses when making decisions on how to power missions, including the timing of when such decisions are made. We also met with officials from NASA's Human Exploration and Operations Mission Directorate to learn whether NASA expects future human exploration missions to use RPS. We interviewed officials at NASA's Glenn Research Center to discuss how, if at all, advancements in RPS and solar technology could affect future demand for

---

[1] National Aeronautics and Space Administration, NASA Space Flight Program and Project Management Requirements, NPR 7120.5D (Sept. 28, 2011).

Pu-238-fueled missions. We also interviewed members of the scientific community and representatives from a commercial entity to obtain their perspectives on the use of RPS for future NASA missions.

To evaluate DOE's progress in meeting NASA's RPS and Pu-238 demand and what challenges, if any, DOE faces in meeting the demand, we reviewed DOE documentation, including DOE's program management guidelines, Pu-238 Supply Project (Supply Project) plans, monthly reports, and RPS and Supply Project management presentations. We reviewed DOE Order 413.3B to evaluate its applicability to the Supply Project.[2] In addition, we reviewed the *Plutonium-238 Production Program Management Plan[3] and the Pu-238 Supply Project - Project Execution Plan[4]* to gain an understanding of the planning, management, and execution of the Supply Project. We obtained and reviewed additional documents associated with the Supply Project, such as DOE's risk management plan and risk register, to determine how DOE is managing new Pu-238 production and the extent to which DOE identifies and mitigates risks. In addition, we reviewed Idaho National Laboratory (INL), Los Alamos National Laboratory (LANL), and Oak Ridge National Laboratory (ORNL) monthly reports to assess periodic updates from each laboratory on the Supply Project and RPS production. We also reviewed presentations from DOE officials who oversee RPS production and the Supply Project. In addition, we interviewed officials from DOE's Office of Nuclear Energy to determine the extent to which DOE is prepared to meet NASA's demand for RPS using Pu-238 for planned and future missions. We also discussed with these officials how DOE prioritizes Pu-238 work within its plutonium strategy and how DOE integrates RPS production with other DOE activities. We interviewed officials at DOE's national laboratories involved in RPS production—INL, LANL, and ORNL—to discuss the Supply Project and RPS production. We conducted site visits to ORNL, the laboratory responsible for the Supply Project, and

---

[2] 3U.S. Department of Energy, Office of Nuclear Energy, Plutonium-238 Production Program Management Plan (Washington, D.C.: September 2015).

[3] U.S. Department of Energy, Office of Nuclear Energy, *Plutonium-238 Production Program Management Plan* (Washington, D.C.: September 2015).

[4] U.S. Department of Energy, Oak Ridge National Laboratory Site Office, Pu-238 Supply Project - Project Execution Plan (April 2016).